CHROMATOGRAPHY: CONCEPTS AND CONTRASTS

CHROMATOGRAPHY: CONCEPTS AND CONTRASTS

James M. Miller

Drew University
Madison, New Jersey

A Wiley-Interscience Publication

JOHN WILEY & SONS

New York / Chichester / Brisbane / Toronto / Singapore

Library of Congress Cataloging-in-Publication Data:

Miller, James M.
    Chromatography: concepts and contrasts.

    "A Wiley-Interscience publication."
    Includes bibliographies and index.
    1. Chromatographic analysis. I. Title.

QD79.C4M55 1987   543'.089   87-16084
ISBN 0-471-84821-2

Printed in the United States of America

10 9 8 7 6 5 4

PREFACE

Chromatography has become the premier technique for separations and analyses. The three most important types of chromatography are gas chromatography (GC), liquid chromatography in columns (LC), and thin layer chromatography (TLC). Because of the large amount of information available about each of them, these individual techniques are often treated separately in monographs. However, they do share a common theoretical base, and the most efficient education of a novice would be a unified study. Furthermore, a scientist working with one of the techniques may need to switch to another one, and that could be more easily accomplished if he/she were acquainted with the theories common to the two techniques.

This monograph attempts a unified approach to chromatography and emphasizes the similarities and differences between the major divisions— GC, LC, and TLC. Thus the title is *Chromatography: Concepts and Contrasts*. In addition to the advantages mentioned above, the unified approach permits the use of one set of terms and symbols which should make learning easier. Another consequence of this approach is that the chapters covering the three main topics (GC, LC, TLC) do not stand alone; rather, they build on the introductory chapters, which cover the common principles. Hopefully, the book is short enough that the novice can afford the time to begin at the beginning and read through to the chapter that covers the information on the topic he/she is seeking.

The unified approach was taken in my earlier book, *Separation Methods in Chemical Analysis*, and some of the material in that book has been repeated. This monograph is more narrow in coverage, of course, and has been updated with the latest information on chromatography. It is an elementary introduction to the topic, but one that is quite comprehensive. In most cases, references are provided for further information, and this book should serve as a good reference text. It includes many practical operating hints. These are presented as evidence of the applicability of the theory and not as hints without rationales. Modern LC is the newest of the three techniques and the one requiring the most description, so Chapter 9 on LC is the longest one. Chapter 11 describes some special techniques, including chiral separations and derivatization. The final chapter, Chapter 12, provides some suggestions for selecting a chromatographic method.

Although the book is introductory and elementary, some background in chemistry will be needed in order to understand the chemical systems described. The better the reader's background in the theory of chemistry, the more meaningful will be the sections that deal with chromatographic theory. A person with a minimal background will probably want to skip lightly over some of the theory, while a chemist may want to delve into the theory more deeply by reading some of the references provided. Only the most basic and necessary information is presented here in order to keep this book as simple as possible.

The book is intended to appeal to a wide audience—academic and industrial scientists, students in both formal coursework and informal private study, and chemists and nonchemists. The academic use is best suited for the undergraduate level, but with additional assignments from the references, it could probably be used at the lower graduate level. In industry it could be used by scientists entering any of the fields of chromatography, or it could serve as an updating for those in the field.

I want to express my appreciation to a large number of people, beginning with the many scientists on whose work I have drawn in the preparation of this monograph. Many of them have taken time to communicate with me privately and I appreciate their kind efforts. Several colleagues have read parts of the manuscript and offered useful suggestions, and I am deeply indebted to them. Thanks also to my students, at Drew and elsewhere, who suffered with early versions of the manuscript and offered helpful criticisms. Special thanks for sabbatical support go to Drew University and Sandoz, Inc.

JAMES M. MILLER

Madison, New Jersey
November 1987

CONTENTS

CHROMATOGRAPHY: CONCEPTS AND CONTRASTS

INTRODUCTION TO CHROMATOGRAPHY 1

Separation methods are an important part of analysis, and chromatography has developed into *the* premier separation technique. Consider the analysis of peppermint oil by gas chromatography shown in Figure 1.1. Twenty-three peaks have been identified, but over 100 peaks have been separated in this chromatogram, which is typical of the high efficiency separations now possible. Chromatography's rapid development can be attributed to its relative simplicity and the successful application of theory to practice.

While developments will continue, thus precluding the writing of a definitive monograph, this work is intended to present a unified view of the current status of chromatography, including the underlying theories and the current state-of-the-art practice.

A BRIEF HISTORY

It can be argued that modern, high performance chromatography began with the publication of Martin and James's paper on gas chromatography in 1952.[1] It is certainly true that their publication on the use of a gas as a mobile phase in the separation of volatile fatty acids initiated the research that has resulted in the wisespread use and popularity of chromatography.

Although chromatography entered a new phase in the early 1950s, the Russian botanist Tswett is generally referred to as the *Father of Chromatography*. His work, published in 1906, described the separation of plant pigments by column liquid chromatography, defined the terms, and demonstrated the technique so well that it was used unchanged for about forty years. The original paper is of significant historical interest and serves as a fitting introduction to a discussion of the concepts of chromatography; fortunately, it has been translated into English and is readily available.[2]

In the period between 1906 and 1952 there were some developments of importance. For example, the techniques of plane chromatography were developed. Earliest was the use of paper as a plane support, but when thin layers of silica gel were introduced as an alternative in the late

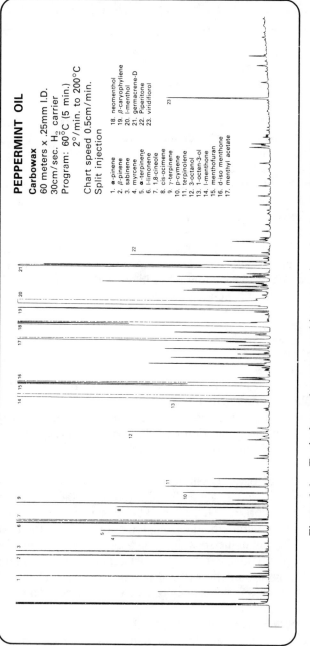

PEPPERMINT OIL

Carbowax
60 meters x .25mm I.D.
30cm/sec. H₂ carrier
Program: 60°C (5 min.)
2°/min. to 200°C

Chart speed 0.5cm/min.
Split injection

1. α-pinene
2. β-pinene
3. sabinene
4. myrcene
5. α-terpinene
6. l-limonene
7. 1,8-cineole
8. cis-ocimene
9. γ-terpinene
10. p-cymene
11. terpinolene
12. 3-octanol
13. 1-octen-3-ol
14. l-menthone
15. menthofuran
16. d-iso menthone
17. menthyl acetate

18. neomenthol
19. β-caryophyliene
20. l-menthol
21. germacrene-D
22. Piperitone
23. viridiflorol

Figure 1.1. Typical gas chromatographic separation showing the high efficiency of this method. Courtesy of J & W Scientific.

1950s, the field of thin layer chromatography (TLC) was born and became so popular that it has largely replaced the older technique.

Column chromatography developments accelerated in the 1940s. Martin and Synge published their Nobel Prize winning paper in which they introduced liquid–liquid (or partition) chromatography and the accompanying theory that became known as the *plate theory*.[3] It is instructive to note that in their important paper, the authors suggested that gas chromatography would be an interesting technique to explore, but no one did so until Martin himself returned to it ten years later. The plate theory was further explored by Craig, who published a paper entitled "Partition Chromatography and Countercurrent Distribution" in 1950.[4] His countercurrent distribution apparatus gained some popularity, but was soon replaced by liquid chromatography for most applications. In 1955 Glueckauf also published a paper titled "The Plate Concept in Column Separations".[5]

As an alternative to the plate theory, the so-called *rate theory*, came into prominence about the same time. The paper that has had the greatest impact was the one published by the Dutch workers van Deemter, Zuiderweg, and Klinkenberg.[6] They described the chromatographic process in terms of kinetics and examined diffusion and mass transfer. The popular *van Deemter plot* resulted. A few years later, Giddings published another paper on this topic,[7] and the rate theory has since become the backbone of chromatographic theory.

The new technique, gas chromatography or GC, was found to be simple and fast and capable of producing separations of volatile materials that were impossible by distillation. Furthermore, the theories were found to be rather accurate in predicting optimal operating conditions, and the theories could be quickly tested. The field exploded! New separations led to new ideas to be tested and vice versa. GC quickly matured.

It was natural to attempt to apply the successful results from GC to the older technique of LC, liquid chromatography. Some of the credit for that transfer of technology belongs to Giddings. In 1963 he published a paper entitled "LC with Operating Conditions Analogous to Those of GC".[8] This set off a revolution in LC that brought it to a level of efficiency similar to that achieved in GC. The acronym HPLC was born. It is not clear if HP originally meant high performance (as it usually does today) or high pressure (which was required to get the superior performance). In either case, HPLC is usually used to distinguish between the new, modern mode of operation as opposed to the old Tswett method, but the American Society for Testing and Materials and others have recommended that the acronym not be used[9]; that recommendation will be followed in this book.

This is only a small glimpse of the historical development of chromatography; it is a fascinating story, and more complete accounts have been published by Ettre.[10]

DEFINITIONS AND CLASSIFICATIONS

Separation

Separation is a familiar word, and it is unlikely that any confusion will arise from its use in reference to chemical analysis. Nevertheless, there is some merit in considering a precise, general definition such as that suggested by Rony[11]:

> Separation is the hypothetical condition where there is complete isolation, by m separate macroscopic regions, of each of the m chemical components which comprise a mixture. In other words, the goal of any separation process is to isolate the m chemical components, in their pure forms, into m separate vessels, such as glass vials or polyethylene bottles.

The adjective *hypothetical* is used for two reasons. In the first place, it is *theoretically* impossible to accomplish the *complete* separation of the components of a mixture, and, in the second place, the separated components often are not actually isolated but rather are detected and their presence recorded (on chart paper).

Consider analytes A and B present together in a homogeneous phase. In order to separate them, a second phase will have to be added (or formed), so that A goes preferentially into one phase while B goes into the other one.

Chromatography

In chromatography, one phase is held immobile or stationary, and the other one is passed over it (the mobile phase). The designations GC and LC refer to the physical state of the mobile phase. Further classifications can be made by naming the mobile and stationary phases; thus we have gas–solid (GS), gas–liquid (GL), liquid–liquid (LL), and liquid–solid (LS) chromatography. More recently, supercritical fluids have been used as mobile phases, and these techniques have been named *supercritical fluid chromatography* (SFC) irrespective of the state of the stationary phase. Other names have also become popular, and Table 1 shows a complete classification scheme. Included in the classification scheme are not only the states of the two phases but also the configuration of the chromato-

TABLE 1 Classification of Chromatographic Techniques

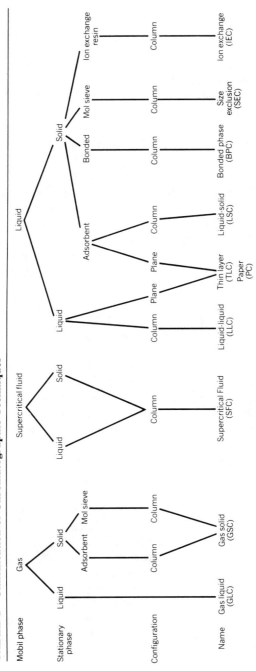

5

graphic *bed*. There are two popular configurations for the bed, a column and a planar surface, and they will be treated separately in this monograph.

An operational definition of chromatography is: chromatography is a separation method composed of two phases, one stationary and one mobile. A mixture of analytes is introduced into the mobile phase and is carried through the system by it. However, as the mobile phase passes over and through the stationary phase, the components of the mixture equilibrate or partition between the two phases, resulting in differential migration rates through the system. Alternatively, we could say that the various components of the mixture are retarded in their passage through the system in proportion to their interaction with (sorption on) the sorbent bed. At any given time, a particular analyte molecule is either in the mobile phase, moving along at its velocity, or in the stationary phase and not moving at all. The sorption–desorption process occurs many times as the molecule moves through the bed, and the time required to do so depends mainly on the proportion of time it is sorbed and held immobile. A separation is effected as the various components emerge from the bed at different times, which are called *retention times*. The process is depicted in Figure 1.2. Since all the molecules of a particular analyte do not behave exactly the same, the peak for that analyte will have a finite width in the chromatogram.

The type of chromatography just described is called *elution* chromatography, since the sample is continually washed or eluted through the system by mobile phase. A less popular form of chromatography is controlled by *displacement*. The sample is pushed through the system by displacing it from the stationary bed with other sample components or with a strong mobile phase. Displacement chromatography is sometimes used in LSC, but it will not be discussed further.

The chromatographic process we have defined is also known as *zonal* or *batch* chromatography because the sample is applied to the system all at once in one narrow zone. By contrast, the sample can be applied continuously during a run; this process is called *frontal analysis*, and it will not be discussed further because of its limited use.

The mode of interaction between the sample components and the two phases can be classified into two types, although many separation processes are combinations of both. If the sample is attracted to the surfaces of the phases, commonly to the surface of a solid stationary phase, the process is called *ad*sorption. Alternatively, if the sample diffuses into the interior of the stationary phase—for example, into the bulk of a stationary liquid—chromatographers call the process *partition*. Actually, *ab*sorption seems to be a better name for this process because we can then speak of

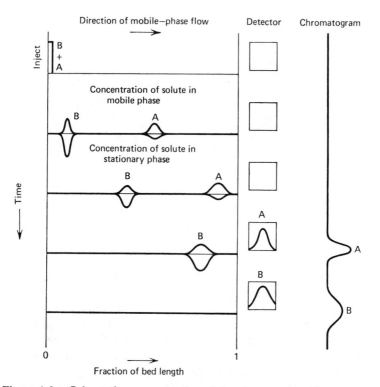

Figure 1.2. Schematic representation of the chromatographic process.

sorption as the general process and add the prefixes *ab* or *ad* when we want to be more specific. For this reason, the terms *ab*sorption and *ad*-sorption will be used in this monograph even though *partition* is usually used by chromatographers to denote the former interaction. A comic illustration of the terms is shown in Figure 1.3.

Chromatographic Symbols

Unfortunately, the terms and symbols used in chromatography are not completely standardized, so it is difficult to choose those to be used in this text. Many different groups, including IUPAC and ASTM, have published recommendations, but differences still exist, partially because GC and LC require different symbols. Within these limitations, the symbols used in this text will be as consistent as possible with those approved by recognized groups, and further details and clarifications can be found in the papers by Ettre, who has summarized the situations in GC[12] and LC.[13]

ABsorption   ADsorption

Figure 1.3. The difference between absorption and adsorption.

Included in Figure 1.2 is a plot of time versus detector signal, which is the common way of presenting chromatographic data and is called a *chromatogram*. Two sample components or peaks are shown in Figure 1.2, and it is clear that they have been separated, which is the objective of chromatography. For purposes of discussion, a more simple chromatogram is given in Figure 1.4, which shows only one sample component plus a small peak representing a nonretained component (which may not always appear). This figure will be used to illustrate some of the chromatographic definitions and symbols.

Most chromatographs are operated with a constant flow (F) of mobile phase unless the flow is intentionally being changed or programmed. Consequently, the x axis of the chromatogram can be labeled as time (t) or

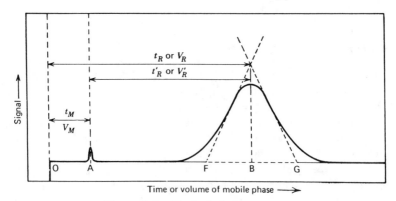

Figure 1.4. Typical chromatogram.

as volume (V) since

$$V = t \times F \tag{1}$$

Thus, V represents the volume of mobile phase that flowed during a specified operating time. If we wish to designate the time required for a component to elute from the chromatographic system, the so-called *retention time*, the symbols t and V are designated with a subscript R.

$$V_R = t_R \times F \tag{2}$$

In making an actual measurement of retention time or retention volume from a chromatogram, one usually measures the *distance* on the chart from the start of the chromatogram to the maximum of the peak of interest. Thus, V_R and t_R can be represented as the distance from 0 to B as indicated in Figure 1.4. Distance on the chart can be converted to time by multiplying it by the chart speed and then to volume by multiplying the time by the flow rate. When constant flow is assumed, retention time and volume can be used interchangeably, and both are shown on the abscissa in Figure 1.4.

Part of the time an analyte spends in the chromatographic system is the time required to go through the "dead" space in the system, and part is caused by the time it spends in the stationary phase, not moving. Thus, the total time or volume can be broken down into two parts:

$$V_R = V_M + KV_S \tag{3}$$

where V_M represents the "dead" space or mobile phase volume, V_S is the stationary phase volume, and K is the partition coefficient. A rigorous derivation of this equation has been published,[12] but the significance of the equation should be obvious without it; that is, the total volume of mobile phase (V_R) required to elute an analyte is composed of two parts: the dead space in the system, which is occupied by mobile phase and through which every analyte must pass (V_M), and the mobile phase which flows while the analyte is held immobile (equal to KV_S). The latter contribution is determined by the length of time the analyte is immobile, which is in turn determined by the amount (volume: V_S) of stationary phase and the tendency of the analyte to sorb in the stationary phase as measured by its partition coefficient K. To recap, the equation shows that there are only two things a given analyte molecule can do: move down the column with the mobile phase, or sorb in or on the stationary phase.

Before looking at the partition coefficient more closely, note the small peak at A in Figure 1.4. It represents the time required for a material to pass through the system without being sorbed, and thus it measures the dead volume or mobile phase volume V_M. In GC, air or methane is often used to provide such a marker; in LC, there is no single simple marker, but a slight shift in the baseline is sometimes observed, depending on the solvent and the detector.

The retention volume (or time) that has the mobile phase volume subtracted out is of interest for theoretical work.

$$V_R - V_M = V_R' = KV_S \qquad (4)$$

It is called the *adjusted* retention volume (or time), and it is designated with a superscript prime. It is also shown in Figure 1.4.

The partition coefficient K is defined as the concentration of an analyte (A) in the stationary phase divided by its concentration in the mobile phase.

$$K = \frac{[A]_S}{[A]_M} \qquad (5)$$

The use of a typical equilibrium constant K in chromatographic theory indicates that the system can be assumed to operate at equilibrium; that is, as the analyte A proceeds through the system at a given temperature, it partitions between the two phases and is retained in the system in proportion to its affinity for the stationary phase. Figure 1.2 is drawn to illustrate this partitioning by showing an analyte zone as two peaks, one in the mobile phase and one in the stationary phase. Analyte A is shown moving faster down the column than analyte B because molecules of A spend less time in the stationary phase, as indicated by the smaller peak in that phase; that is, A has a smaller partition coefficient K than B. Even though this is a dynamic system, the ideal operating parameters are such that the system is not far from equilibrium and the use of a partition coefficient is valid.

Equation (5) is greatly simplified; the true thermodynamic partition coefficient would be the quotient of analyte *activities*, not *concentrations*. Furthermore, the equation assumes that analyte A is present in only one form (one molecular structure or ion). When this is not realized in practice, a more complex equilibrium constant must be used.

Another assumption that is usually not stated is that the analytes do not interact with each other; that is, molecules of analyte A pass through the chromatographic system as though no other analytes were present. This assumption is reasonable because of the low concentrations at which analytes are present and because they are increasingly separated from each other as they pass through the system.

In making use of the partition coefficient in chromatography, it is useful to break it down into its two parts—β, the phase volume ratio and k, the partition ratio:

$$K = \beta k \tag{6}$$

where

$$\beta = \frac{V_M}{V_S} \tag{7}$$

and

$$k = \frac{(\text{mass of A})_S}{(\text{mass of A})_M} \tag{8}$$

The partition ratio is also called the *capacity factor*, and another common symbol is k'.

By combining Eqs. (4), (6), and (7), Eq. (9) can be derived. It serves as another definition of k and can easily be measured (Figure 1.4).

$$k = \frac{V_R'}{V_M} = \frac{t_R'}{t_M} \tag{9}$$

Another chromatographic parameter is the retention ratio R_R. It is the relative average speed v of an analyte through a chromatographic system compared to the average mobile phase speed or velocity u.

$$R_R = \frac{v}{u} \tag{10}$$

It will always be equal to or less than 1, and it expresses the fractional rate at which an analyte is moving. It also represents the fraction of molecules of a given analyte in the mobile phase at any given time and, alternatively, the fraction of time an average analyte molecule spends in the mobile phase as it travels through the system.

Each of the velocities in Eq. (10) can be defined and measured according to the length L of the chromatographic system.

$$v = \frac{L}{t_R} \tag{11}$$

and

$$u = \frac{L}{t_M} \tag{12}$$

A combination of these three equations shows that

$$R_R = \frac{t_M}{t_R} = \frac{V_M}{V_R} \tag{13}$$

Furthermore, substituting Eq. (13) into Eq. (3)

$$R_R = \frac{V_M}{V_M + KV_S} \tag{14}$$

$$= \frac{1}{1 + k} \tag{15}$$

When these relationships are compared with the analogous ones for liquid–liquid extractions, it is noted that retention ratio in chromatography is similar in concept to the *fraction extracted* in extractions.

The Normal Distribution

The shapes of all the peaks in the chromatograms shown thus far are symmetrical, approximating the normal or Gaussian shape. This is the ideal shape, but it is not always achieved in practice. Theoretically, the Gaussian shape is closely approached if the analyte has undergone a sufficiently large number of sorptions and desorptions, as is the case for most peaks with partition ratios of about 1 or greater. In fact, an asymmetrical peak is usually evidence that some undesirable interaction is taking place in the system.

For all the theoretical discussions in this book the Gaussian shape will be assumed, thus allowing us to draw some conclusions about the peaks and to use some standard nomenclature and symbols to describe chromatographic peaks. Remember that a peak represents the frequency distribution of all molecules of a particular analyte as they move as a group through the system or as they are detected as they exit from the system. The "average" molecule is found at the center of the distribution, at the position of the peak maximum, and it is this "average" position that is used to characterize the particular analyte.

The familiar equation for the normal distribution is

$$y = \frac{1}{\sigma\sqrt{2\pi}}\exp\left[-\frac{1}{2}\left(\frac{x - \bar{x}}{\sigma}\right)^2 \right] \tag{16}$$

where y is the dependent variable, x is the independent variable, $\bar{x}$ is the average of a large number of x's, and σ is the standard deviation. In the analysis of peaks resulting from a separation, it will be most useful to express the variable x in units of standard deviation. Hence, for our purposes, Eq. (16) can be written as

$$y = \frac{1}{\sqrt{2\pi}}\exp\left(-\frac{\sigma^2}{2}\right) = 0.3989\ \exp\left(-\frac{\sigma^2}{2}\right) \tag{17}$$

This equation is plotted in Figure 1.5. Also shown as broken lines are tangents to the points of inflection. Where they intersect the baseline,

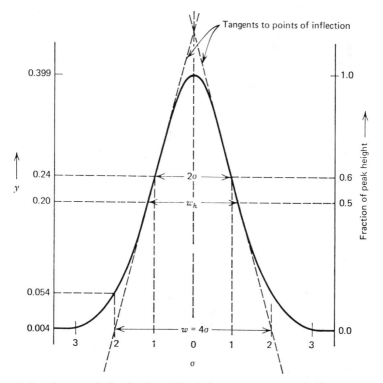

Figure 1.5. A normal distribution. The inflection point occurs at 0.607 of the peak height. The quantity w_h is the width at 0.500 of the peak height and corresponds to 2.354σ.

they cut off the distance w known as the *peak width* (at the base). It can be seen from the figure that w has a value of 4σ ($\pm 2\sigma$). Consequently σ, the standard deviation, is also called the *quarter peak width* (at the base). Note also that the width at 60.7% of the peak height is 2σ ($\pm 1\sigma$), and at 50% of the peak height it is 2.354σ. The latter is called the *peak width at half height*, w_h.

Several measures of asymmetry, such as the one shown in Figure 1.6, have been devised by chromatographers for asymmetric peaks. One is called the *asymmetric ratio* or *tailing factor*, TF;

$$ \text{TF} = \frac{b}{a} \tag{18} $$

where a and b are measured at 10% of the height of the peak as shown. A symmetrical peak will have a value of 1 and tailed peaks a value of greater than 1. A *fronted* peak (one with a *leading edge* and $a > b$) will have a value less than 1.

Other Terms

Plate Number. The most common measure of the efficiency of a chromatographic system is the plate number, n. Because the concept originated from the analogy with distillation, it was originally called the *number of theoretical plates* contained in the chromatographic column (system). This is not a useful analogy because a chromatographic column does not

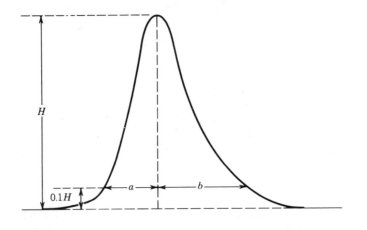

Figure 1.6. Figure used to define asymmetric ratio or tailing factor.

contain "plates," but the terminology has remained, and the original definition has persisted:

$$n = \left(\frac{t_R}{\sigma}\right)^2 \tag{19}$$

The parameters t_R and σ must be measured in the same units, so the number n is dimensionless (see Figure 1.7). For Gaussian peaks, we can express σ in terms of peak width since we know the relationships as just presented in the last section. For example, the base width w is equal to 4σ, so n becomes

$$n = \left(\frac{4t_R}{w}\right)^2 = 16\left(\frac{t_R}{w}\right)^2 \tag{20}$$

Similarly, w_h corresponds to 2.354σ, and n equals

$$n = 5.54\left(\frac{t_R}{w_h}\right)^2 \tag{21}$$

Refer to Figure 1.5 for a summary of these relationships.

If the peak is not symmetrical, different values will be calculated for n because the width measurements will not follow the predicted Gaussian distribution. In general, for asymmetrical peaks, n increases the higher up on the peak the width is measured.

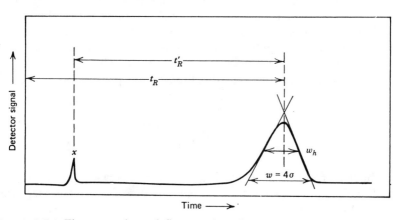

Figure 1.7. Figure used to define n, the plate number (x = nonretained component).

One method of calculating the plate number that does not depend on symmetry uses statistical moments. A summary of moment analysis is:

Zeroth moment measures peak area.

First moment measures peak mean and hence t_R.

Second moment measures peak variance and hence σ.

Third moment measures peak asymmetry.

The only practical way to make these measurements is with a microcomputer and appropriate data collection software. Further details are provided in several recent papers.[14]

The plate number is a measure of the relative peak broadening (w) that has occurred while the analyte passed through the system (in time t_R). As we will see, peaks broaden (w increases) as the retention time increases, which is the reason peak width alone is not sufficient to specify the efficiency of the system. Furthermore, according to theory, n should increase slightly as k increases, but the effect is usually not large, so column efficiencies are usually given without regard to the particular analyte and its k value.

A related measure of system efficiency is the *effective* plate number, n_{eff}:

$$n_{eff} = 16\left(\frac{t_R'}{w}\right)^2 \qquad (22)$$

in which the adjusted retention time is used instead of the retention time (see Figure 1.7). We have already noted that the adjusted retention time has more theoretical significance than the retention time, and consequently the effective plate number is often a better parameter for use in comparing chromatographic columns, particularly comparing packed columns to unpacked (open tubular) columns. By combining equations already presented, it can be shown that the effective plate number is related to n by

$$n_{eff} = n\left(\frac{k}{k+1}\right)^2 \qquad (23)$$

The effective plate number will increase as k increases, and it will approach n at high k values where V_M is no longer of significant size compared to V_R.

As already mentioned, the number of theoretical plates is a term also used in distillation, but it is important to note that comparisons of efficiencies of the two techniques *cannot* be made by comparing plate numbers. It takes more chromatographic plates to achieve a given separation by GC than it does to achieve the same separation by distillation.[15]

Plate Height. The plate number will depend on the length of the column, making comparisons among columns difficult unless all are of the same length. Another related parameter that removes this dependence is the plate height, H.

$$H = \frac{L}{n} \tag{24}$$

L is the total column length, so H can be thought of as the length of column that contains one plate. Clearly, n and H are inverse to each other and H is a measure of efficiency that is independent of total length. It has the units of length, usually centimeters or millimeters, and, like n, it originated in distillation. In fact, it is sometimes referred to by the full name used in distillation—HETP, or *height equivalent to a theoretical plate*. But, again, it should be emphasized that the column does not contain plates, and the use of these two terms results from their historical development.

Resolution. A better measure of the efficiency of a chromatographic system is resolution, R_s. It defines the degree of separation of two analytes or peaks,

$$R_s = \frac{d}{(w_A + w_B)/2} = \frac{2d}{w_A + w_B} \tag{25}$$

where d is the distance between the peak maxima and w is the width of each peak at the base as shown in Figure 1.8. The larger the resolution, the better the separation; a value of 1.0 is shown in the figure representing about 98% resolution.

This value can be easily understood if it is assumed that the widths of the two peaks are the same: $w_A = w_B$. In this case, $R_s = 2d/2w = d/w$. Since the tangents to the peaks (dotted lines) are just touching, and since $w = 4\sigma$ for each peak, d must also equal 4σ (2σ from A plus 2σ from B) and $R_s = 4\sigma/4\sigma = 1$.

For simplicity, Eq. (25) is often reduced to

$$R_s = \frac{d}{w_B} \tag{26}$$

since the peak widths are usually the same. A resolution of about 1.5 is necessary for complete separation, and values less than 1.0 represent poorer separations than that shown in Figure 1.8.

Strictly speaking, Eq. (26) is only valid when both peaks have the same height, as is shown in Figure 1.8. A theoretical approach to the calculation of resolution for peaks of unequal height in GC was published by Glueckauf,[16] but it has been shown to be in error.[17] A better, practical approach has been provided by Snyder.[18] His paper includes computer-drawn pairs of peaks for a variety of resolutions and peak height ratios from 1 : 1 to 128 : 1. Some of them are shown in Figure 1.9. Snyder's recommendation is to compare a given separation with those he has provided and to choose the best match, assigning that value of resolution to the chromatogram in question. This practice has become accepted and seems adequate. Other possibilities have been suggested by Carle,[19] Karger[20] (for peaks with different widths), and Bly[21] (for size exclusion chromatograms).

Peak Capacity. A final measure of column efficiency is the peak capacity, or the number of peaks that can be resolved (R_s = 1) by a given system in a given time (t_R or V_R). Giddings[22] introduced it in 1967 and used it to compare the potential separating capabilities of the various chromatographic modes. Figure 1.10 shows a series of analytes represented by triangles and a peak capacity of six, and Table 2 lists the values he obtained for three chromatographic systems—GC, LC, and SEC (size exclusion chromatography).

A similar concept was suggested by Kaiser,[23] who calls his parameter a *separation number* (SN), or *Trennzahl* (TZ) in German. It is the number of analytes that can be resolved between two consecutive members of the paraffin homologous series x and $x + 1$. It can be calculated by Eq. (27).

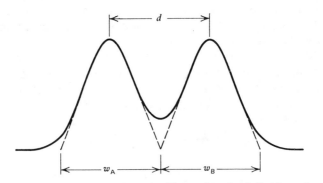

Figure 1.8. Two nearly resolved peaks illustrating the definition of resolution.

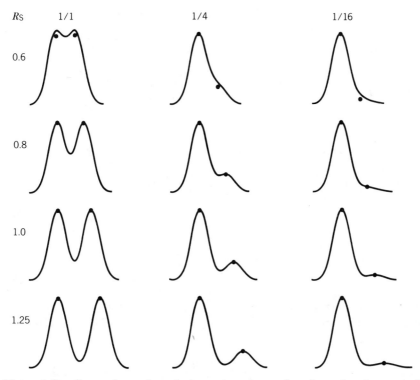

Figure 1.9. Comparison of resolution values for peaks of equal and nonequal heights. Reproduced from the *Journal of Chromatographic Science* by permission of Preston Publications, Inc.

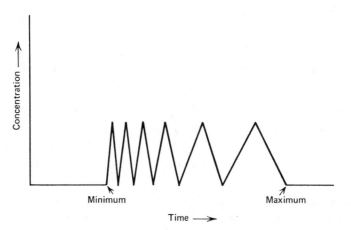

Figure 1.10. A hypothetical separation illustrating the concept of peak capacity.

19

TABLE 2   Calculated Peak Capacities for Different Chromatographic Techniques[a]

Plate No.	Peak Capacity		
	GC	LC	SEC
100	11	7	3
400	21	13	5
1000	33	20	7
2500	51	31	11
10000	101	61	21

[a] Reprinted with permission from J. C. Giddings, *Anal. Chem.* **1967**, 39, 1027. Copyright 1967, American Chemical Society.

$$TZ = \frac{t_{R(x + 1)} - t_{R(x)}}{w_{h(x + 1)} + w_{h(x)}} - 1 \qquad (27)$$

This term is very similar to resolution, which is equal to $\Delta t_R/1.7w_h$, when defined with similar symbols. Substituting it into Eq. (27), we get

$$TZ = 0.85R_s - 1 \qquad (28)$$

which can be rearranged to

$$R_s = 1.177TZ + 1.177 \qquad (29)$$

Thus, baseline resolution of 1.5 is equivalent to a TZ value of 0.275, and a TZ value of 1.0 requires an R_s of 2.35.

TZ numbers can be used for programmed temperature operation in GC and gradient elution in LC, conditions under which plate numbers are not considered to be valid. Regardless of the operating conditions, a TZ number of 12, for example, means that 12 analytes eluting adjacent to each other ($R_s = 1$) can be resolved between two consecutive paraffins in that region of the chromatogram. Figure 1.11, taken from Freeman,[24] compares n, n_{eff}, and TZ for some paraffins on an open tubular GC column. For further discussion of peak capacity, see the paper by Ettre.[25]

Column Selectivity. Separations can be effected by differences in partition coefficients as well as by the efficiencies of the columns in which they

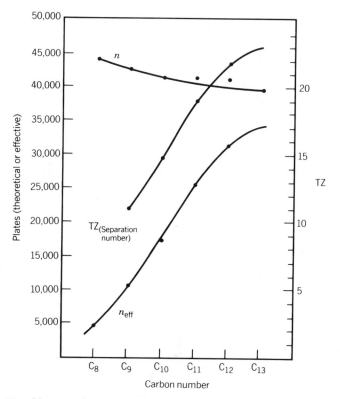

Figure 1.11. Plate number and separation number vs. carbon number for 10 m × 0.25 mm OT column at 100°C. © Copyright 1981 Hewlett-Packard. Reproduced with permission.

are run. Thus, another useful measure of the separability of two analytes is the ratio of their partition coefficients on a given column. This ratio is called the *column selectivity* or, more generally, the *separation factor*, α:

$$\alpha = \frac{K_B}{K_A} = \frac{(V'_R)_B}{(V'_R)_A} \tag{30}$$

It is usually defined so that its numerical value is >1.0, which would mean that analyte B is retained longer on the system than analyte A. Chapter 4 contains more discussion of the three parameters α, *n*, and *k*.

TABLE 3 Summary of Important Chromatographic Equations

1. $(K)_A = \dfrac{[A]_S}{[A]_M}$

2. $K = k\,\beta$

3. $\beta = \dfrac{V_M}{V_S}$

4. $\alpha = \dfrac{K_B}{K_A} = \dfrac{(V_R')_B}{(V_R')_A}$

5. $V_R = V_M + K\,V_S$

6. $V_N = K\,V_S$

7. $k = \dfrac{(W_A)_S}{(W_A)_M} = \dfrac{V_R'}{V_M} = \left(\dfrac{V_R}{V_M}\right) - 1 = \dfrac{1 - R_R}{R_R} = \left(\dfrac{1}{R_R}\right) - 1$

8. $R_R = \dfrac{V_M}{V_R} = \dfrac{\bar{v}}{\bar{u}}$

   $\quad = \dfrac{V_M}{V_M + K\,V_S}$

9. $R_F = \dfrac{S_{analyte}}{S_{solvent\ front}}$

10. $V_R = V_M(1 + k) = \dfrac{L}{\bar{u}}(1 + k) = n(1 + k)\dfrac{H}{\bar{u}}$

11. $(1 - R_R) = \dfrac{k}{k + 1}$

12. $R_R(1 - R_R) = \dfrac{k}{(k + 1)^2}$

13. $n = 16\left(\dfrac{t_R}{w}\right)^2 = \left(\dfrac{t_R}{\sigma}\right)^2 = 5.54\left(\dfrac{t_R}{w_h}\right)^2$

14. $H = \dfrac{L}{n}$

15. $n_{eff} = \left(\dfrac{k}{k + 1}\right)^2 n$

16. $h = \dfrac{H}{d_p}$

17. $v = u\,\dfrac{d_p}{D_M}$

18. $R_S = \dfrac{2d}{w_A + w_B}$

19. $TZ = \dfrac{t_{R(x + 1)} - t_{R(x)}}{w_{h(x + 1)} + w_{h(x)}} - 1$

SUMMARY

A large number of terms, symbols, and equations were given in this chapter. The equations are gathered together in Table 3, along with a few others that will be introduced in later chapters. As commonly used, some symbols are slightly different for GC and LC, but this should not diminish the value of the table. The Appendix contains a list of symbols and acronyms used in this text.

The Appendix also contains two chromatograms and all the operating conditions needed to make calculations of the most important parameters. It is recommended that these calculations be performed in order to gain a better understanding of their meaning. A review published by Meyer[26] can also be consulted for a variety of LC calculations.

REFERENCES

1. A. T. James and A. J. P. Martin, *Biochem. J.* **1952**, *50*, 679.
2. H. H. Strain and J. Sherma, *J. Chem. Educ.* **1967**, *44*, 238.
3. A. J. P. Martin and R. L. M. Synge, *Biochem. J.* **1941**, *35*, 1358.
4. L. C. Craig, *Anal. Chem.* **1950**, *22*, 1346 – 1352.
5. E. Glueckauf, *Trans. Farad. Soc.* **1955**, *51*, 34–44.
6. J. J. van Deemter, F. J. Zuiderweg, and A. Klinkenberg, *Chem. Eng. Sci.* **1956**, *5*, 271.
7. J. C. Giddings, *J. Chem. Phys.* **1959**, *31*, 1462.
8. J. C. Giddings, *Anal. Chem.* **1963**, *35*, 2215.
9. ASTM, *Standard Practice for Liquid Chromatography Terms and Relationships*, ANSI/ASTM E 682-79, American Society for Testing and Materials, Philadelphia, 1979.
10. L. S. Ettre and A. Zlatkis (eds.), *75 Years of Chromatography—A Historical Dialogue*, Vol. 17, J. Chromatogr. Library, Elsevier, Amsterdam, 1979; L. S. Ettre and C. Horvath, *Anal. Chem.* **1975**, *47*, 422A; L. S. Ettre, *Anal. Chem.* **1971**, *43*(14), 20A; *Am. Lab.* **1978**, *10*(10), 85, (11), 120.
11. P. R. Rony, *Separ. Sci.* **1968**, *3*, 239.
12. L. S. Ettre, *J. Chromatogr.* **1979**, *165*, 235.
13. L. S. Ettre, *J. Chromatogr.* **1981**, *220*, 29.
14. D. A. Dezaro, T. R. Floyd, T. V. Raglione, and R. A. Hartwick, *Chromatogr. Forum* **1986**, *1*(1), 34; B. A. Bidlingmeyer and F. V. Warren, Jr., *Anal. Chem.* **1984**, *56*, 1583A; J. P. Foley and J. G. Dorsey, *J. Chromatogr. Sci.* **1984**, *22*, 40; W. W. Yau, *Anal. Chem.* **1977**, *49*, 395.

15. B. L. Karger, L. R. Snyder, and C. Horvath, *An Introduction to Separation Science*, Wiley, New York, 1973, p. 166.

16. E. Glueckauf, *Trans. Faraday Soc.* **1955**, *51*, 34.

17. S. H. Tang and W. E. Harris, *Anal. Chem.* **1973**, *45*, 1979.

18. L. R. Snyder, *J. Chromatogr. Sci.* **1972**, *10*, 200.

19. G. C. Carle, *Anal. Chem.* **1972**, *44*, 1905.

20. B. L. Karger, *J. Gas Chromatogr.* **1967**, *5*, 161.

21. D. D. Bly, *J. Polymer Sci. Part C*, **1968**, *21*, 13.

22. J. C. Giddings, *Anal. Chem.* **1967**, *39*, 1027.

23. R. Kaiser, *Chromatographie in der Gasphase*, 2d ed., Vol. 2, Bibliographisches Institut, Mannheim, West Germany, 1966, pp. 47–48.

24. R. R. Freeman (ed.), *High Resolution Gas Chromatography*, 2d ed., Hewlett-Packard, 1981.

25. L. S. Ettre, *Chromatographia* **1975**, *8*, 291.

26. V. R. Meyer, *J. Chromatogr.* **1985**, *334*, 197.

SELECTED BIBLIOGRAPHY

Giddings, J. C., *Dynamics of Chromatography*, Part 1, Dekker, New York, 1965.

Heftmann, E. (ed.), *Chromatography: Fundamentals and Applications of Chromatographic and Electrophoretic Methods*, 2 vols., 4th ed., Elsevier, Amsterdam, 1983. Part A has individual chapters on fundamentals and techniques, and Part B has 15 chapters on separate applications.

Karger, B. L., L. R. Snyder, and C. Horvath, *An Introduction to Separation Science*, Wiley-Interscience, New York, 1973.

Miller, J. M., *Separation Methods in Chemical Analysis*, Wiley-Interscience, New York, 1975.

Zweig, G. and J. Sherma (eds.), *CRC Handbook of Chromatography*, CRC Press, Boca Raton, Fla. This is a continuing multivolume set including the following:
G. Zweig and J. Sherma, *General Data and Principles*, 2 vols., 1973.
M. Qureshi, *Inorganics*, 1986.
S. Blackburn, *Peptides*, 1986.
J. C. Touchstone, *Steroids*, 1986.
J. M. Follweiler and J. Sherma, *Pesticides and Related Organic Chemicals*, 1984.
C. J. Coscia, *Terpenoids*, 1984.
H. K. Mangold, *Lipids*, 2 vols., 1984.
S. Blackburn, *Amino Acids and Amines*, 1983.
C. G. Smith, N. E. Skelly, C. D. Chow, and R. A. Solomon, *Polymers*, 1982.
T. Hanai, *Phenols and Organic Acids*, 1982.
S. H. Churms, *Carbohydrates*, 1982.
R. N. Gupta and I. Sunshine, *Drugs*, 2 vols., 1981.

KINETIC PROCESSES IN CHROMATOGRAPHY 2

As was discussed in the preceding chapter the chromatographic process is controlled by equilibrium, and as a consequence its retention parameters are related to thermodynamic partition coefficients or equilibrium constants. However, chromatography is dynamic, and concentration gradients exist in an analyte zone as it passes through the chromatographic bed or column. Those gradients result in diffusion, a kinetic process. Also involved in the dynamics of chromatography is the process of mass transfer. These topics are the subject of this chapter. First, however, let us look at the stationary phases used in chromatography.

CONFIGURATIONS OF THE STATIONARY PHASE

The stationary phase can be a liquid or a solid. If it is a liquid, it can be coated directly on the inside walls of a capillary tube (column), or it can be coated on an inert *solid support* and be handled like a solid. In effect, then, there are three stationary phase configurations: in the first type, a solid (with or without stationary liquid) is packed into a column; in the second type, a solid is coated on the surface of a flat, plane material like glass (TLC), and in the third type, a liquid is coated on the inside wall of an open tube (OT).

Packed columns and thin layer plates have similar characteristics in that they contain a solid support or packing that is not present in OT columns. We need to examine more closely the nature of packed beds and will use a packed column as an example.

Chromatographic packings have small particle sizes and are tightly and uniformly packed into columns, as shown in Figure 2.1. The particles can be spherical or irregular, porous (like sponges as shown in Figure 2.1) or nonporous (like glass beads). If the stationary phase is a liquid, its coating on the solid support should be thin and uniform like that shown in Figure 2.2. If it is not, pools of stationary liquid phase can be formed in the pores of the particles (Figure 2.3a) or between the particles (Figure 2.3b). Such nonuniform films are undesirable.

The total volume in the column, V_T, is made up of three parts:

$$V_T = V_{ss} + V_S + V_M \tag{1}$$

V_{ss} = volume occupied by the solid support
V_S = volume occupied by the stationary phase
V_M = volume occupied by the mobile phase

If the solid is nonporous, V_M is the space between the particles (the *interparticle volume*), but if it is porous, V_M includes both the interparticle volume and the internal volume of the particles. (In some instances, the size of the pores in the solid may be too small to admit the analyte molecules, and thus the internal portions of the solid may not be accessible to the sample, but that is a complication we will ignore at present.)

These two types of particle give rise to two definitions of porosity.

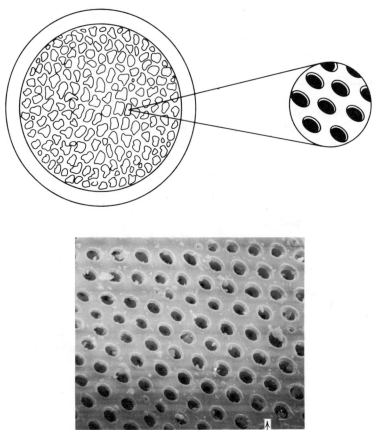

×4,700   1 μm

Figure 2.1. Magnified view of a packed column. Higher magnification is of Chromosorb P at ×4700.

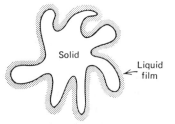

Figure 2.2. Uniform distribution of liquid on solid support.

The total porosity, ϵ_T, is the fraction of *dead space* in the column

$$\epsilon_T = \frac{V_M}{V_T} \tag{2}$$

Even in a column that is packed very tightly with small particles, this porosity can be as high as 0.80 to 0.84 for a porous material; that is, a tightly packed bed of porous particles is still about 80% empty!

The other measure of porosity, ϵ_I, is only that fraction of volume between the particles

$$\epsilon_I = \frac{\text{interparticle volume}}{V_T} \tag{3}$$

It varies with the tightness of the packing and has a typical value of 0.40 to 0.45. Nonporous, or solid-core, solids have only one porosity, of course.

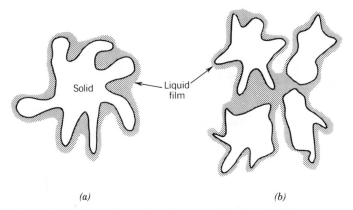

(a)   *(b)*

Figure 2.3. Nonuniform distribution of liquid on solid support.

THE RATE THEORY

The brief historical development in the last chapter noted that the early theoretical papers described chromatography in terms similar to distillation or extraction and were known as the *plate theory*. Useful as it may have been in the development of chromatography, the plate theory is of little value in modern chromatography and has been replaced by the *rate theory*. Any of the early books on gas chromatography can be consulted for a discussion of the plate theory, and Giddings[1] has written a good historical summary of the concurrent development of the plate and rate theories.

The Original Van Deemter Theory

The important paper by van Deemter, Klinkenberg, and Zuiderweg[2] described the kinetic factors involved in GC as an analyte passed through a packed bed. It identified three effects that contribute to zone spreading during this process: eddy diffusion or the multipath effect (the A term), longitudinal molecular diffusion (the B term), and mass transfer in the stationary phase (the C term). The broadening of a zone was expressed in terms of H, the plate height, and was described as a function of the average linear mobile phase velocity, u. Thus, the classical *van Deemter* equation is

$$H = A + \frac{B}{u} + Cu \qquad (4)$$

For GC, a plot of H versus u gives a curve like the one shown in Figure 2.4 and has become known as a *van Deemter* plot.

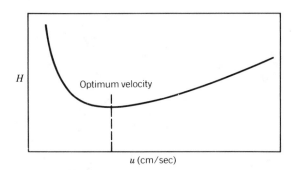

Figure 2.4. Typical plot of rate equation (van Deemter plot).

Many modifications and refinements were proposed to improve the utility of this equation for different situations. Golay[3] showed that the A term was not needed for open tubular columns in GC, and Giddings[4] proposed a complicated coupled term in his extensive treatment of the theory. Huber[5] and others noted that a fourth term was necessary for LC to account for mass transfer in the liquid mobile phase. The exact terms in these variations of the rate theory are given in Table 1 mainly for their historical value.

A Modern View of the Rate Theory

Hawkes[6] has evaluated the various rate equations and presented a modern summary. He suggests that the rate equation should take the form

$$H = \frac{B}{u} + Cu \tag{5}$$

$$= \frac{B}{u} + (C_S + C_M)u \tag{6}$$

$$= \frac{2\psi D_M}{u} + q\frac{k}{(1 + k)^2}\frac{d_f^2}{D_L}u + \frac{fn(d_p^2, d_c^2, end)}{D_M}u \tag{7}$$

where the C term is split into two parts—one for mass transfer in the stationary phase (C_S) and the other for mass transfer in the mobile phase (C_M). The latter includes the original A term, which is no longer a separate term. Let us use this equation to take a closer look at the contributions to zone spreading.

The B term, representing longitudinal molecular diffusion, is the least controversial and is essentially the same as originally proposed by van Deemter. Molecules will diffuse from the region of high concentration (the center of the zone) to the region of lower concentration in proportion to the diffusion coefficient D_M, according to Fick's law. This effect is shown in Figure 2.5 as a function of time; as the time increases from t_1 to t_2 to t_3, the zone spreads and its maximum is lowered, resulting in zone broadening as the analyte proceeds through the system. Some typical diffusion coefficients are given in Table 2.

An obstruction factor ψ is included in the numerator of this term to account for the obstruction to free diffusion caused by packed beds. Some typical values of ψ are given in Table 3. The faster the mobile phase moves, the less time the zone is in the column, the less time there is for diffusion, and the lower is the broadening due to diffusion. Consequently, this term

TABLE 1 Comparison of Rate Equations

Reference	H_{ed}	H_{lmd}	H_{mts}	H_{mtm}	Other
van Deemter	$2\lambda d_p$	$\dfrac{2\psi D_M}{\bar{u}}$	$\dfrac{8}{\pi^2}\dfrac{k}{(1+k)^2 D_S} d_f^2 \bar{u}$	—	—
Extended van Deemter	$2\lambda d_p$	$\dfrac{2\psi D_M}{\bar{u}}$	$\dfrac{k}{q(1+k)^2}\dfrac{d_f^2 \bar{u}}{D_S}$	$\dfrac{\omega d_p^2 \bar{u}}{D_M}$	—
Golay	—	$\dfrac{2D_M}{\bar{u}}$	$\dfrac{k}{q(1+k)^2}\dfrac{d_f^2 \bar{u}}{D_S}$	$\dfrac{1+6k+11k^2}{24(1+k)^2}\dfrac{d_p^2 \bar{u}}{D_M}$	—
Giddings	—	$\dfrac{2\psi D_M}{\bar{u}}+\dfrac{2\psi_S D_S(1-R_R)}{\bar{u}R_R}$	$\dfrac{k}{q(1+k)^2}\dfrac{d_f^2 \bar{u}}{D_S}$	$\dfrac{\omega d_p^2 \bar{u}}{D_M}$	$\left(\dfrac{1}{2\lambda d_p}+\dfrac{1}{H_{mtm}}\right)^{-1}$
Huber	$\dfrac{2\lambda d_p}{1+\lambda_2(D_M/\bar{u}d_p)^{1/2}}$	$\dfrac{a_d D_M}{\bar{u}}$	$a_S\dfrac{k}{(1+k)^2}\dfrac{d_p^2 \bar{u}}{D_S}$	$a_M\left(\dfrac{k}{1+k}\right)^2\dfrac{d_p^{3/2}(\bar{u})^{1/2}}{D_M^{2/3}}$	—

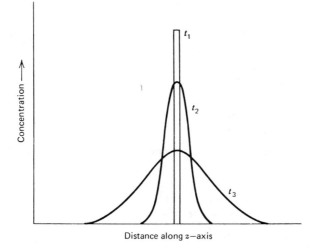

Figure 2.5. Zone widening due to diffusion. Three times are shown with $t_3 > t_2 > t_1$.

must be divided by the mobile phase velocity, and the contribution to zone broadening by the B term is important only when the mobile phase velocity is low, as shown in Figure 2.4. In many chromatographic operations, the mobile phase velocity is sufficiently large that the B term is not of major importance.

The A term, which was known as the *eddy diffusion* effect, does not exist in Hawkes's equation as a separate term. The effect is better called a *multipath term*; it is flow-dependent, contrary to the original van Deemter equation, and is included in the C term. An exact equation would be very complicated and is inappropriate because of the negligible contri-

TABLE 2 Typical Diffusion Coefficients

System	D (cm²/sec)	Temperature (°C)
Liquids		
Glucose–water	0.52×10^{-5}	15
CCl$_4$–methanol	1.7×10^{-5}	15
Acetic acid–benzene	1.92×10^{-5}	14
Gases		
Argon–n-octane	0.059	30
Nitrogen–n-octane	0.073	30
Helium–n-hexane	0.574	144
Helium–methanol	1.032	150

TABLE 3 Approximate Obstruction Factors[a]

Chromatographic Bed	ψ
Glass beads	0.6–0.7
Chromosorb W	0.74
Chromosorb P	0.46
Paper	0.7–0.9

[a] From Giddings.[1]

bution to zone broadening at normal chromatographic mobile phase velocities.

The C term describes mass transfer effects in the broadening process. They can best be understood by reference to Figure 2.6, which depicts an analyte zone distributed between the two phases. The upper zone

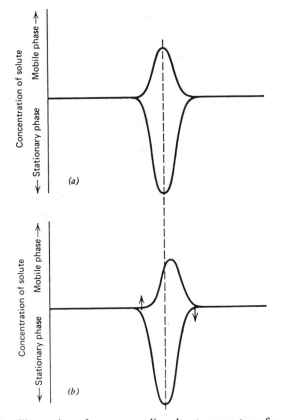

Figure 2.6. Illustration of zone spreading due to mass transfer ($K = 2.0$).

represents the distribution of analyte in the mobile phase, and the lower one represents the distribution in the stationary phase. A partition coefficient of 2 is shown so that twice as much analyte is in the stationary phase as is in the mobile phase. As shown in Figure 2.6a, the analyte is at equilibrium. The situation an instant later is shown in Figure 2.6b, where the flow of the mobile phase has carried some analyte with it down the column and ahead of the zone mean, thus widening the overall zone. In order to restore the system to equilibrium, some analyte in the leading edge of the mobile zone must sorb on the stationary phase, and some analyte in the trailing edge of the stationary zone must desorb into the mobile phase. This mass transfer takes a finite time and results in zone broadening. The faster the mass transfer occurs, the less will be the effect, so chromatographic phases are designed with this in mind.

A fast-moving mobile phase will make the time needed for mass transfer of analyte between the two phases more important because the zone is swept swiftly ahead of the places where equilibration is slow, so the C term is multiplied by the velocity u. Furthermore, the term can be broken down into two parts—one for mass transfer in the stationary phase C_S, and the other in the mobile phase C_M. The combined term is $(C_S + C_M)u$.

If the stationary phase is a solid, the mass transfer process is adsorption and desorption on its surface. This process is very fast and usually does not contribute significantly to the C_S-term.

If the stationary phase is a liquid, the original stationary phase mass transfer term of van Deemter is basically correct. For example, Eq. (7) shows that H is proportional to the square of the thickness of the stationary liquid film, d_f, and inversely proportional to the diffusion coefficient in the liquid phase, D_L. Thus a good column will have a small film thickness, and the stationary liquid will be chosen for its high diffusion coefficient.

The coefficient q was originally given as $8/\pi^2$ by van Deemter, but a better value for a uniform liquid film is probably $\frac{2}{3}$. Values of q for other configurations are given in Table 4.

The other part of the stationary phase mass transfer term is the ratio $k/(1 + k)^2$, which is also equal to $R_R(1 - R_R)$, an alternative ratio some-

TABLE 4 Approximate Configuration Factors

Type of Bed	q
Uniform liquid film	$\frac{2}{3}$
Paper	$\frac{1}{2}$
Ion-exchange resin	$\frac{2}{15}$

times used in the equation. As k increases, R_R decreases and so does the ratio, as shown in Table 5. Since a small value is desirable, this term suggests the use of a large k. However, as k increases so does the time of analysis, and this is a consideration to be discussed later. Also, beyond a k value of about 10, very little is gained by further increases in k, which is a second reason for not using extremely large k values.

Equation (7) showed that the term for mobile phase mass transfer was an indefinite combination of a several factors, including particle size d_P, column diameter d_c, and the diffusion coefficient D_M in the mobile phase. It includes the old concept of *eddy diffusion*, as noted earlier, and it is significantly important only in LC. It is indeed complex, and only the most important parameters will be discussed here; the paper by Hawkes[6] should be consulted for further details.

Zone spreading of analyte molecules in the mobile phase will depend upon the square of the diameter of the particles (in a packed column) for several reasons. One relates to the existence of stagnant pools of mobile phase within the pores of a porous support. Analyte molecules that diffuse into these pores will get "trapped" and fall behind the analyte zone. Secondly, the analyte must travel from particle to particle in the flow stream as it undergoes the chromatographic process of sorption and desorption. The larger the particles, the larger this distance between sorptions, the greater the width of the flow stream through which analyte molecules diffuse, and the greater the width of the analyte zone. Both of these factors depend upon the particle diameter d_P; in fact, the dependence turns out to be equal to the square of d_P. This effect is diminished by fast diffusion of the analyte in the mobile phase, so the diffusion coefficient D_M is in the denominator.

The column diameter d_c is also a factor for both packed and open tubular columns. In both cases, the zone spreading is a function of the square of the diameter. Other parameters that enter into this term include: the diameter of the coil of the column, which is usually avoided by using uncoiled (straight) columns in LC; end effects that cause broadening as the sample enters and leaves the column; and the multipath effect, which arises from the fact that each analyte molecule can take a different path

TABLE 5   Comparison of k and R_R

k	R_R	$R_R(1 - R_R)$
1	0.5	0.25
4	0.2	0.16
9	0.1	0.09
49	0.02	0.0196

through the packed bed, thus spending more or less time traversing the column and producing some zone spreading as a result. All these effects, like those that give rise to zone spreading in the stationary phase, will increase as the mobile phase velocity increases, so this term is also multiplied by the mobile phase velocity u.

When the rate equation is applied to GC, allowance must be made for the fact that the mobile phase (a gas) is compressible and the linear velocity u is not a constant but increases throughout the column. This effect will be discussed further in Chapter 5.

A Redefinition of H

What is H anyway? The original interpretation, taken from distillation theory, was *height equivalent to a theoretical plate*, or HETP. We have seen that this concept was inadequate, and the preceding discussion of the van Deemter equation has presented it as a measure of the extent of spreading of an analyte zone as it passes through a column. Thus, a more appropriate term might be *column dispersivity*. In fact, another, independent approach to the theory of chromatography defines H as

$$H = \frac{\sigma^2}{L} \tag{8}$$

where σ^2 is the variance or distribution of analyte molecules about their mean in the analyte zone, as we have seen earlier, and L refers to the length (or distance) of movement of the analyte zone. Thus H is truly the dispersion of an analyte per unit length migrated. It is a superior concept and can be related to the actual diffusion process via the Einstein equation

$$\sigma^2 = 2Dt \tag{9}$$

where D is the diffusion coefficient and t is the time of diffusion.

Further elaboration of Eq. (8) can be found in Chapter 4, where slightly different forms will be used for column methods and planar methods. For the present, it should be noted that L does not necessarily refer to column length, even though that is the definition usually used in this text.

Practical Consequences of the Rate Theory

In Gas Chromatography. The rate equation for GC tells us that our packed column should have thin films on the solid support; d_f should be small. This is achieved by using supports with large surface areas and small amounts of stationary phase (low percentage of liquid phase). The lower limit for liquid phase percentage is that point at which the solid support is not fully covered. This limit is reached at about 0.3% (by weight) for a typical diatomaceous earth solid support. Since d_f is in the C_S term and is of most importance at the large gas flow rates that are preferred in GC, it is probably the most important consideration in designing a GC column.

Other parameters in this important C term are D_L, the diffusion coefficient in the liquid phase, q, and the ratio $k/(1 + k)$.[2] The diffusion coefficient should be large, but often this choice cannot be exercised because the liquid phase is chosen for selectivity reasons. Previously, it had been thought that higher diffusion coefficients would be found in stationary phases of low viscosity. This is true only for small molecules; for polymers of the type used in GC, the diffusion coefficient is virtually independent of viscosity.[7] Of course, these polymers cannot be used below their glass temperatures, so low viscosity polymers may be required for low temperature GC column operation. The configuration factor q is determined by the type of bed and is $\frac{2}{3}$ for uniform films preferred in GC. The k ratio should be relatively large, as we have seen.

These same principles apply to open tubular (OT) columns, except that the thin liquid film is deposited directly on the walls of the column rather than on the solid support. Typical OT columns have an inside diameter of 0.25 mm and a film thickness of 0.25 μm.

Mass transfer in the mobile phase is an important contributer to zone broadening for packed columns. To minimize its effects, columns should be tightly and uniformly packed with small particles. Obviously, this is not a factor with open tubular columns, since they have no packing.

The effect of carrier gas is shown in Figure 2.7, which gives typical van Deemter plots for the common carrier gases helium, hydrogen, and nitrogen in OT columns. As predicted by the B term of the rate equation, nitrogen, which has a higher molecular weight and smaller diffusion coefficients, is more efficient and gives the smallest dispersivity H. This advantage is usually offset by the better performance of helium and hydrogen at higher velocities where the B term is not important, as shown in the figure; for fast analyses, helium and hydrogen are definitely preferred.

Hydrogen has been gaining in popularity for use with open tubular columns because it has several advantages. Its optimum velocity is around 35 to 40 cm/sec, even higher than that for helium, permitting fast analyses. Its curve is the flattest, so changes in velocity that can occur during programmed temperature operation will not result in significant changes in H. Its viscosity is the lowest of the three gases, permitting its use at lower inlet pressures. In addition, hydrogen is less expensive than helium, especially for the higher purities. However, its use requires that the col-

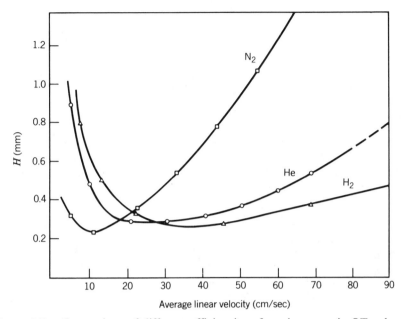

Figure 2.7. Comparison of different efficiencies of carrier gases in OT column GC. © *Research & Development.*

umn oven be monitored for leaks for safety's sake, and that the flame detector be modified, since hydrogen is now supplied through the column and not as a separate fuel.

In Liquid Chromatography. Molecular diffusion, the B term, is seldom of importance in LC because operating velocities are usually well above those at which its effects would be noticed. On the other hand, both C terms are important. As a consequence, plots of the rate equation usually differ somewhat from GC plots, as shown in Figure 2.8.

Optimum operating conditions for LC, as determined by the C terms, are similar to those for GC: d_f and d_P should be small and D_L should be large. Also, as noted earlier, LC columns are usually straight, rather than coiled, to prevent the race track effect. Other minor factors in the C_M term were discussed earlier.

Figure 2.8 shows that the rate equation can be considered to be a straight line if a short segment of the plot is considered. Snyder[8] has shown that the straight line can be approximated by

$$H = D(u)^{0.4} \tag{10}$$

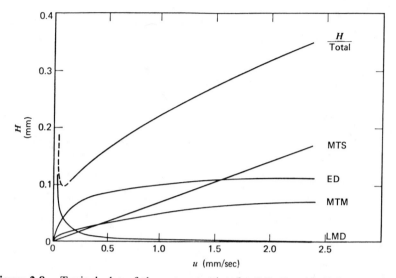

Figure 2.8. Typical plot of the rate equation for LC. Reprinted from J. F. K. Huber, *J. Chromatog. Sci.* **1979**, *7*, 85 by permission of Preston Publications, Inc.

for particles larger than 10 μm where D is a constant:

$$D = 18(d_P)^{0.8} \tag{11}$$

However, for smaller particles, down to 3 μm, he and others have found that a more general equation[9] is needed:

$$H = Au^{0.33} + \frac{B}{u} + Cu \tag{12}$$

His paper[10] should be consulted for recommendations on selecting the best experimental conditions for small particle columns. He considers decreasing the column pressure (and flow rate), increasing the column length, and combinations of these. His scheme is based on a *reduced* form of Eq. (12), which is presented in Chapter 5.

REFERENCES

1. J. C. Giddings, *Dynamics of Chromatography*, Part 1, Dekker, New York, 1965, pp. 13–25.

2. J. J. van Deemter, F. J. Zuiderweg, and A. Klinkenberg, *Chem. Eng. Sci.* **1956**, *5*, 271.

3. M. J. E. Golay in *Gas Chromatography 1958*, D. H. Desty (ed.), Butterworths, London, 1958, p. 36.

4. Giddings, *Dynamics of Chromatography*, pp. 40–65.

5. J. F. K. Huber and J. A. R. J. Hulsman, *Anal. Chem.* **1967**, *38*, 305.

6. S. J. Hawkes, *J. Chem. Educ.* **1983**, *60*, 393–398.

7. S. J. Hawkes, *Anal. Chem.* **1986**, *58*, 1886.

8. L. R. Snyder, *J. Chromatogr. Sci.* **1969**, *7*, 352.

9. E. Grushka, L. R. Snyder, and J. Knox, *J. Chromatogr. Sci.* **1975**, *12*, 25.

10. L. R. Snyder, *J. Chromatogr. Sci.* **1977**, *15*, 441.

PHYSICAL FORCES AND
MOLECULAR INTERACTIONS

3

The process of selecting a chromatographic system for a given separation will be facilitated if we have some understanding of the forces that are operating in the system, giving rise to the separation. The magnitude of these forces is expressed in the numerical values of the partition coefficients, which were shown in Chapter 1 to be a part of the fundamental theory of chromatography. In this chapter, we will first consider some fundamentals regarding forces and then we will see how these concepts have been applied to chromatographic separations.

INTERMOLECULAR AND INTERIONIC FORCES

Introduction

The terms *adsorption* and *absorption* (or *partition*) were introduced in Chapter 1 along with the definition of the partition coefficient K. Now we want to explore the forces that give rise to a particular partition coefficient and also see what effect they have on peak shape.

Sorption Isotherms. If the partition coefficient K is a true constant, a plot of $[A]_S$ versus $[A]_M$ should be a straight line, as shown in Figure 3.1a. We say that the system is *linear*, and we refer to such a graph as an *isotherm*, since the data are presented at a constant temperature. The chromatographic peak resulting from such a relationship is symmetrical and has the desired Gaussian distribution we have discussed earlier, as shown in Figure 3.2a. However, when the concentration of analyte gets large enough, deviations from linearity are usually observed, as shown by the deviation from the dashed line in Figure 3.1a. Small analyte concentrations are preferred in chromatography so that linearity is obtained.

In systems that operate mainly by adsorption, it is also common to find isotherms that deviate from linearity. The resulting isotherm, known as a *Langmuir isotherm* (Figure 3.1b), arises when the concentration of analyte in the mobile phase exceeds the linear level; that is, the stationary phase does not have the capacity to sorb all the analyte that would be

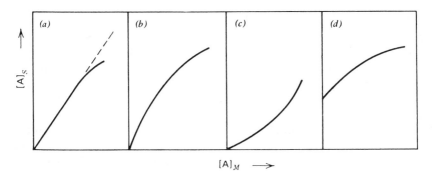

Figure 3.1. A comparison of four types of isotherm: (a) linear; (b) Langmuir; (c) anti-Langmuir; (d) chemisorption.

required for a linear system. This type of behavior is common in GS and LS chromatography and can result from sorption on a surface that has heterogeneous energy sites. The first analyte molecules adsorb on the most active sites, and additional molecules (at the higher concentration levels) see a surface that is much less active, and they do not adsorb as much as the earlier molecules did, resulting in nonlinearity. Sites capable of hydrogen bonding with the analyte are often the cause of this effect. The peak shape that results from a Langmuir isotherm is shown in Figure 3.2b; the skewed shape is usually referred to as tailing.

The opposite type of isotherm (*anti-Langmuir*) is shown in Figure 3.1c and its resulting peak shape in Figure 3.2c. In this situation, the analyte molecules, which are the first to adsorb, facilitate the sorption of additional molecules. For example, in LC when the planar molecule phenol adsorbs on alumina, the phenyl rings extend out from the surface, as

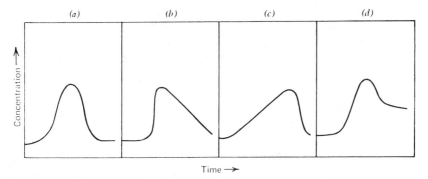

Figure 3.2. Peak shapes corresponding to the isotherms in Figure 3.1: (a) linear; (b) Langmuir; (c) anti-Langmuir; (d) chemisorption.

shown in Figure 3.3. Additional molecules will not be attracted by the alumina surface that attracted the first molecules but by the phenyl rings, which will facilitate increased solubility. In GC this effect occurs when the temperature is too low or the system is overloaded (too large a sample).

Another consequence of nonlinear sorption is that the retention time of a given analyte will be a function of its size, as shown in Figure 3.4. In the case of tailing (Langmuir isotherms), the retention time decreases as the sample size increases, and the opposite is true for fronting (anti-Langmuir) peaks. Clearly this is a problem if one wants to make identifications based on retention times or if one's computer is looking in a specified time "window" for a given peak.

The fourth isotherm, Figure 3.1d, arises from chemisorption: a chemical reaction rather than a physical force. The resulting chromatogram shown in Figure 3.2d is characterized by a baseline that very slowly (or never) returns to its original position (zero) because the desorption is so slow. In fact, there are cases, especially in GC, where a basic chemical such as an amine reacts completely with the acidic surface of the stationary phase and is never seen at all. Many repeated injections are necessary to react all of the acidic sites before any amine can be eluted from the column.

Let us now look at the types of forces that are present in chromatographic systems.

Ionic Interactions

Ions are favored in aqueous solutions and are important mainly in LC, although some GC work has been reported with molten ionic salts for stationary liquid phases.[1] Forces between ions of like charge are repulsive and between ions of different charge are attractive, according to Coulomb's law. They are relatively long-range and strong forces. The main illustration of ionic forces in LC is in ion exchange where the sample is at least partly ionized and the stationary phase contains ionic sites.

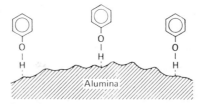

Figure 3.3. Representation of the adsorption of phenol on alumina.

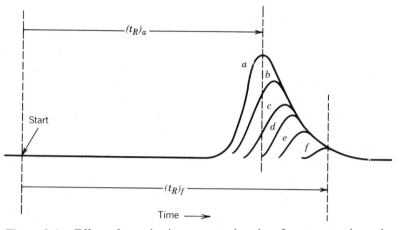

Figure 3.4. Effect of sample size on retention time for asymmetric peak.

Ions are also attracted and repulsed by the polar ends of a dipolar molecule. These forces can be referred to as *ion–dipole forces*. In addition to ion exchange chromatography, they are also important in aqueous LC systems where a polar stationary phase can interact with ionic solutes.

Van der Waals Forces

Three types of weaker forces have been identified and together are referred to as *van der Waals forces*. They are listed in Table 1 according to the type of interaction: dipole–dipole, dipole–induced dipole, and induced dipole–induced dipole. They are more often called by the names listed in the second column or by the names of the men who first described them, as listed in the third column.

Orientation forces are easy to understand by analogy with magnetic forces. The opposite poles attract even though random molecular motion keeps the attraction small. Similarly, the induction forces can be viewed as similar to a magnet attracting nonmagnetic iron. The induction of a dipole depends on the polarizability of the nonpolar molecule; large molecules with easily deformed electronic clouds have large polarizabilities.

TABLE 1 Classification of Nonionic Intermolecular van der Waals Forces

Interaction	Name	Investigator
Dipole–dipole	Orientation	Keesom (1912)
Dipole–induced dipole	Induction	Debye (1920)
Induced dipole–induced dipole	Dispersion	London (1930)

Dispersion forces cannot be explained by the magnetic analogy nor by conventional electrostatics. They are weak forces that exist even in monoatomic gases that are symmetrical and nonpolar. It is believed that at any particular instant this symmetry is somewhat distorted due to the motion (and position) of the electrons of a given atom, which produces a momentary polarity. This momentary polarity can attract and be attracted by a similar polarity in a neighboring atom or molecule in such a way as to produce a net attraction. As we have already seen, such inductions depend on the polarizability of the molecule or atom. Dispersion forces will always be possible between molecules, but they are the *only* forces between nonpolar hydrocarbons such as the alkanes. For this reason alkanes are often chosen as the ideal molecules for study or for use as standards. An example of a chromatographic separation in which the only forces are dispersion forces would be the GC separation of alkanes on squalane, a branched paraffin.

Table 2 contains some typical values for these three van der Waals forces for a few atoms and simple molecules. Beginning with the noble gases that are nonpolar and have no induction or orientation forces, we can see that the polarizability increases as the size increases and that the dispersion forces also increase. The symmetrical molecule hydrogen has no permanent dipole and only a small dispersion force between its molecules.

The other molecules in Table 2 have permanent dipoles and exhibit all three forces. Note that in the series of hydrogen halides, the dipole moment decreases as the size and polarizability increase; hence orientation and induction forces decrease as dispersion forces increase in the series.

TABLE 2   Relative Magnitude of Intermolecular Forces

Substance	Polarizability $\rho(\text{Å}^3)$	Dipole Moment μ (debyes)	Molecular Volume (mL/mole)	Energy (kcal/pair; $\text{Å}^6 \times 10^{23}$)		
				Keesom	Debye	London
He	0.21	0	26.0	0	0	3.6
Ar	1.63	0	23.5	0	0	165
Xe	4.0	0	27.2	0	0	650
H_2	0.81	0	24.7	0	0	27
CO	1.99	0.12	26.7	0.008	0.13	160
HCl	2.63	1.03	23.4	44	13	265
HBr	3.58	0.78	29.6	15	10	440
HI	5.4	0.38	35.7	0.84	4	880
NH_3	2.24	1.50	20.7	200	24	165
H_2O	1.48	1.84	18.0	450	24	110

Hydrogen Bonding

Hydrogen bonds are formed between molecules containing a hydrogen atom bonded to an electronegative atom like oxygen or nitrogen. Such is the case in alcohols, amines, and water, all of which can both donate and receive a hydrogen atom forming hydrogen bonds. Other molecules such as ethers, aldehydes, ketones, and esters can only accept protons—they have no hydrogens to donate; they can form hydrogen bonds only with hydrogen donors such as alcohols.

These "forces" are strong enough (about 5 kcal/mol) to be considered weak *bonds*—that is, chemical reaction products. The distinction between a force and a bond is not sharp, of course, and hydrogen bonds are classified here with the other forces. However, hydrogen bonds are directional in space, and in this regard they are like bonds and not like the other forces just described, which are randomly oriented in space. As we have already seen, the strength of hydrogen bonds makes them very important in separation processes and may lead to nonsymmetrical peak shapes. The surfaces of many supports used in chromatography contain hydroxyl groups that can form hydrogen bonds and give the surface an undesirable heterogeneity. Later we will discuss the attempts that have been made to remove these groups by derivatization.

Charge Transfer

Finally, there is a group of specific interactions in which two molecules or ions combine by transferring an electron from one to the other. The process is called *charge-transfer*[2] and a *charge-transfer complex* is formed from the attractive forces produced. One of the most common examples in chromatography is the complex formed between Ag^+ ions and olefins, which has been used to separate olefins from paraffins.[3] A thorough discussion of the general principles of charge-transfer complexes and their uses in GC has been published.[4]

SIZE EXCLUSION—SIEVING

Although not a "force," *sieving* is another mechanism by which separations can be achieved in chromatography. Probably *sieving* is not the best term to use, but it does denote that separations are made on the basis of the sizes of the sample molecules. In fact, in their most common form, chromatographic separations based on size are achieved by controlling the size of the pores in the stationary phase so that some (small) molecules will be able to enter the pores while others (the large ones) cannot. The large molecules are excluded, which is why this process is called *size exclusion*. Molecules of intermediate size will be partially excluded from the pores and can be separated from each other based on the fractional extent of their exclusion.

In GC this process has been used to separate fixed gases such as hydrogen, oxygen, nitrogen, methane, carbon monoxide, ethane, carbon dioxide, and ethylene[5] and it has been called *molecular sieve* chromatography. The sieves are natural zeolites or synthetic materials of which the alkali metal aluminosilicates are typical. Table 3 lists the pore sizes of some commercial sieves. Newer sieves have been especially prepared from carbon; Figure 3.5 shows a separation on a typical one, carbosieve II-S.

In LC the main application has been the characterization of polymers using synthetically prepared stationary phases of varying pore sizes. The technique was formerly known by as many as eleven different names, including *gel filtration* and *gel permeation chromatography*.[6] Some confusion may result from the use of these names, which are no longer recommended.

It must be remembered that the solids used as stationary phases in this technique also contain some polar functional groups, and adsorption by analytes does occur in many cases. Thus, the mechanism of their action is often more complex than a simple size exclusion.

Another separation mechanism that depends primarily on the size of the solute molecules involves the formation of inclusion compounds, clathrates or adducts.[7] Stable complexes are formed whereby one (called the *guest*) is trapped inside the other (called the *host*). Van de Waals forces are present and help stabilize the complexes, and in some cases hydrogen bonds are involved in forming the cages. These methods are not widely used, but they can produce some exceptional separations. A general discussion can be found in references 8 and 9, and some typical applications are in references 10 (GC), 11 (TLC), and 12 (LC).

TABLE 3   Some Molecular Sieves

Type	Pore Diameter (Å)
3A	3
4A	4
5A	5
10X	8
13X	9–10

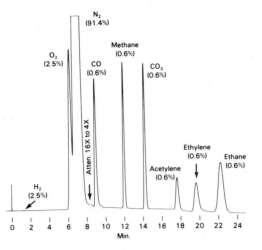

Figure 3.5. Programmed temperature GC separation of gases on Carbosieve S-II column. 100/120 Carbosieve S-II, 10′ × 1/8″ SS; col. temp.: hold 7 min. at 35°C, then to 225°C at 32°C/min., Flow rate: 30 mL/min. He. Reprinted with permission from the catalog of Supelco, Inc., Bellefonte, PA.

SOME MODELS

In order to achieve a separation using chromatography, the components must be retained by the stationary phase to differing degrees. In general, then, the stationary phase is chosen so that sample components will be attracted to it; that is, the stationary phase should interact with the sample, and the modes of interaction are primarily the forces we have been discussing. In simplest terms, we can be guided by the organic chemists' slogan "like dissolves like," which means that the sample will interact with the stationary phase to the greatest extent if it is *like* the stationary phase. In GC, for example, one might choose a polyglycol stationary phase to separate alcohols; both the stationary phase and the analytes have hydroxy groups that can form hydrogen bonds as well as van der Waals attractions and produce different retention times for the analytes, thus separating them.

There are many examples where this simple generalization holds, as will become obvious as the examples in this book are examined. However, a stationary phase of *opposite* polarity might be best to separate analytes that differ little in functionality. For example, the relatively nonpolar xylene isomers are best separated on a polar column that effects the separation by highlighting the slight differences in polarity among the isomers (see Figure 3.6). In addition, there are many sample mixtures that contain analytes with a variety of functional groups and polarities, and the choice of a stationary phase is more complicated. Chapters 8 and 9 contain more information about the choice of stationary and mobile phases, but many separation systems are chosen by trial and error when theory is inadequate to handle a complex system. In the following section we will explore some of the attempts that have been made to describe the nature of the system

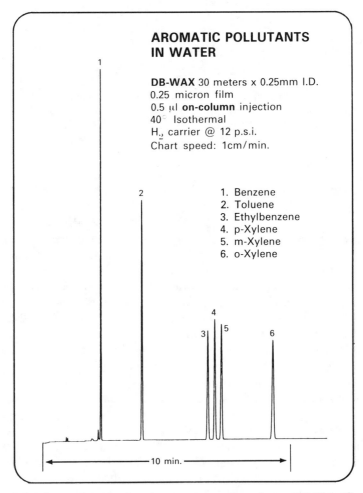

Figure 3.6.   Separation of xylene isomers on a polar column, DB-Wax. Courtesy of J & W Scientific.

from a theoretical point of view. Remember that our chromatographic systems can be represented as solutions (GLC, LLC) and/or as surface adsorptions (GSC, LSC).

Hildebrand's Solubility Parameter

An *ideal* solution is one in which the interactions between solute and solvent molecules are the same as those for the pure solvent and pure solute. It follows Raoult's law and has no heat of mixing, no entropy of mixing, and no change in volume on mixing. Very few solutions fit this model, so other, less ideal models have been proposed.

Hildebrand has defined a *regular* solution[13] as one in which deviations from ideality are attributed only to the enthalpy of mixing; the intermolecular forces are limited to dispersion forces. The equation that defines his model is

$$\Delta \mathcal{H}_{mix} = \Delta \mathcal{G}_{mix} = \mathcal{R}T\ln \gamma_A = \phi_1^2 V_A(\delta_1 - \delta_A)^2 \tag{1}$$

where ϕ_1 is the volume fraction of the solvent, V_A is the molar volume of solute A, γ_A is the activity coefficient of solute A, and δ_1 and δ_A are the so-called *solubility parameters* for the solvent and the solute respectively. This new parameter, the Hildebrand solubility parameter, is a measure of the internal pressure of a substance, defined as

$$\delta = \left(\frac{\Delta \mathcal{E}_{vap}}{\overline{V}}\right)^{1/2} \tag{2}$$

where $\Delta \mathcal{E}_{vap}$ is the molar evaporation energy and $\overline{V}$ is the molal volume of the liquid. Clearly the solubility parameter is a measure of the forces between molecules; it is the square root of the energy of vaporization per milliliter. Table 4 contains some solvents listed according to their increasing solubility parameters.

Chromatographers have tried to use the solubility parameter to predict chromatographic retention in GC[14] and in LC.[15] As originally defined, it is applicable only to nonpolar molecules, but modifications have been suggested to adapt it to fit more complicated systems.[16,17]

Snyder's Solvent Parameter

In contrast with the absorption (partition) model just described, Snyder[18] has provided a model of adsorption in LSC. From his very exhaustive and comprehensive study has come an equation that summarizes the in-

TABLE 4   Some Solubility Parameters δ and Solvent
Parameters ϵ^0

Compound	δ	ϵ^0 (Alumina)
n-Pentane	7.1	0.00
n-Hexane	7.3	0.01
Diethyl ether	7.4	0.38
Cyclohexane	8.2	0.04
n-Propyl chloride	8.5	0.30
Carbon tetrachloride	8.6	0.18
m-Xylene	8.8	—
Ethyl acetate	8.9	0.58
Benzene	9.2	0.32
Chloroform	9.2	0.40
Dibutylphthalate	9.3	—
Propyl nitrile	9.96	—
Acetone	10.0	0.56
Isopropanol	11.4	0.82
Acetonitrile	12.1	0.65
Methanol	14.5	0.95
Polyacrylonitrile	15.4	—

teractions in LSC:

$$\log K = \log V_a + E_a(S^0 - A_s\epsilon^0) \qquad (3)$$

where K is the partition coefficient for a given solute, V_a is the *adsorbent surface volume* or the volume of an adsorbed monolayer of mobile phase liquid, E_a is the average surface activity of the solid, S^0 is the adsorption energy of a solute, A_s is the area of solid required by the adsorbed solute, and ϵ^0 is the Snyder eluent strength parameter.

In comparing the effects of two different mobile liquids on a given solute and a given solid stationary phase, it can be shown that the relative partition coefficients are related to each other according to Eq. (4):

$$\log\frac{(K)_1}{(K)_2} = E_a A_s(\epsilon_2^0 - \epsilon_1^0) \qquad (4)$$

The ratio of retention volumes in the two solvents depends on their difference in eluent strength $(\epsilon_2^0 - \epsilon_1^0)$, since E_a and A_s are constant. Thus, if one has tabulated values for the eluent strength parameter ϵ^0, one can predict the effect on retention of changing ϵ^0. This concept forms one basis for the selection of mobile phases in LSC.

Snyder has provided a large number of ϵ^0 values for solutes run on alumina, and some are included in Table 4. By definition, pentane has a value of zero, and increasing values indicate increasing polarity. The larger the value of ϵ^0, the less strongly adsorbed is a given solute from that solvent.

When ϵ^0 values are applied to other adsorbents, they should be multiplied by the relative E_a values suggested by Snyder:

$$\epsilon^0 \text{ (silica)} = 0.77\epsilon^0 \text{ (alumina)} \tag{5}$$

$$\epsilon^0 \text{ (magnesia)} = 0.58\epsilon^0 \text{ (alumina)} \tag{6}$$

$$\epsilon^0 \text{ (Florisil)} = 0.52\epsilon^0 \text{ (alumina)} \tag{7}$$

While the Hildebrand and Snyder parameters are useful in predicting chromatographic behavior, they are not measures of the exact same concept, and their values are not completely consistent. There are other ways to investigate intermolecular forces. One fairly successful way has been to use a few selected solutes as probes. By observing the degree of retention of these probes in a given chromatographic system, one can get a measure of the forces that caused the probe to be retained. For example, if a hydrocarbon is run in a given chromatographic system, its retention will be determined mainly by dispersion forces and perhaps induction forces, depending on the system. Other, more polar probes can be used to measure other forces. This approach has been used in both GC and LC and is discussed later, in the chapters on GC and LC.

REFERENCES

1. F. Pacholec and C. F. Poole, *Chromatographia* **1983**, *17*, 370–374.
2. S. P. McGlynn, *Chem. Rev.* **1958**, *58*, 1113.
3. B. L. Karger, *Anal. Chem.* **1967**, *39*(8), 24A.
4. R. J. Laub and R. L. Pecsok, *J. Chromatogr.* **1975**, *113*, 47.
5. See, for example, H. P. Burchfield and E. E. Storrs, *Biochemical Applications of Gas Chromatography*, Academic Press, New York, 1962, p. 185.
6. D. M. W. Anderson, F. C. M. Dea, and A. Hendrie, *Talanta* **1971**, *18*, 365.
7. M. Baron, in *Physical Methods in Chemical Analysis*, Vol. 4, W. G. Berl (ed.), Academic Press, New York, 1961, p. 223.
8. E. C. Makin, in *New Developments in Separation Methods*, E. Grushka (ed.), Dekker, New York, 1976.
9. E. Smolkova-Keulemansova, *J. Chromatogr.* **1980**, *184*, 347–361; **1982**, *251*, 17.

10. A. C. Bhattacharyya and A. Bhattacharjee, *Anal. Chem.* **1969**, *41*, 2055.
11. V. M. Bhatnagar and A. Liberti, *J. Chromatogr.* **1965**, *18*, 177.
12. J. Zukowski, D. Sybilska, and J. Jurczak, *Anal. Chem.* **1985**, *57*, 2215–2219.
13. J. H. Hildebrand and R. L. Scott, *Regular Solutions*, Prentice-Hall, Englewood Cliffs, N.J., 1962.
14. L. Rohrschneider, *J. Gas Chromatogr.* **1968**, *6*, 5.
15. P. J. Schoenmakers, H. A. H. Billiet, and L. deGalan, *Chromatographia* **1982**, *15*, 205.
16. R. A. Keller, B. L. Karger, and L. R. Snyder, in *Gas Chromatography*, N. Stock and S. G. Perry (eds.), Institute of Petroleum, London, 1970.
17. D. E. Martire and D. C. Locke, *Anal. Chem.* **1971**, *43*, 68.
18. L. R. Snyder, *Principles of Adsorption Chromatography*, Dekker, New York, 1968.

OPTIMIZATION AND THE ACHIEVEMENT OF SEPARATION

4

In the previous chapters we have examined the two factors that must be considered to understand how separations occur. One is the kinetic factor that describes how analyte molecules spread into an increasingly wide zone during their transport through the chromatographic bed. The other is the thermodynamic factor that explains the interactions between analyte and the chromatographic phases resulting in differential sorption or retention in the bed. In this chapter we will combine these two factors and see how a separation is effected. For simplicity, the discussion will be limited to column chromatography.

KINETICS AND ZONE BROADENING

The best definition of zone broadening was given in Chapter 2 as

$$H = \frac{\sigma^2}{L} \tag{1}$$

which expresses the analyte dispersivity per length of column. In that chapter some caution was expressed regarding the interpretation of L; it is not necessarily the length of the column. If it were, all peaks eluting from a column would have the same peak width (4σ) since

$$\sigma = \sqrt{HL} \tag{2}$$

and H and L are constants.

Perhaps it is obvious that the *meaning* of Eq. (1) is that peak width would depend on retention time t_R, since retention time expresses the total time (or number of opportunities) available for zone broadening. This interpretation of L can be shown by combining a few basic equations. The definition of H is

$$H = \frac{L}{n} \tag{3}$$

where L *is* the column length. Since n is defined as

$$n = \left(\frac{t_R}{\sigma}\right)^2 \tag{4}$$

and if t_R and σ are in the same units (time), the substitution of Eq. (4) into Eq. (3) gives

$$H = \frac{L\sigma^2}{t_R^2} \tag{5}$$

In order to equate Eq. (5) with Eq. (1), two different definitions of H, one has to assume that $L = t_R$. Thus, in column chromatography, the interpretation of L is that it represents the retention time, and the definition of σ that is most useful is

$$\sigma = \sqrt{H t_R} \tag{6}$$

from which we conclude that peak width is proportional to the *square root* of the retention time. This relationship correctly describes the increase in peak width that is observed as the retention time increases.

If a given analyte is run on a longer column under the same operating conditions, its retention time will increase in proportion to the column length. Consequently, we can also state that peak width is proportional to the square root of the column length. Based on that fact alone, one might conclude that separations would be difficult for analytes with long retention times on long columns since the peaks would be too wide. However, we need to examine the consequences of thermodynamics on zone migration before coming to a conclusion.

THERMODYNAMICS AND ZONE MIGRATION

From the thermodynamic relationships presented in Chapter 1, we know that

$$t_R = \frac{t_M}{R_R} = t_M(1 + k) \tag{7}$$

$$= t_M\left(1 + \frac{K}{\beta}\right) \tag{8}$$

For a given column (constant t_M and β), the retention time of an analyte will depend upon its partition coefficient K; specifically, retention time will increase linearly with K.

If we now consider two analytes, A and B, which we wish to separate, we can know their relative retention times from their respective partition coefficients. Thus, on a given column, their peak maxima will be separated by a distance d:

$$d = (t_R)_B - (t_R)_A \tag{9}$$

$$= \left(\frac{t_M}{\beta}\right)(K_B - K_A) \tag{10}$$

On a longer column their retention times will be increased and so will their distance of separation; that is, the separation distance is directly proportional to the column length:

$$d \propto L \tag{11}$$

This increasing distance of separation between peaks makes it possible to get a separation.

THE ACHIEVEMENT OF SEPARATION

We have seen that peaks get wider in proportion to the *square root* of column length but that two peaks are separated in direct proportion to the column length. Since the resolution R_S between these two peaks is given approximately by

$$R_S = \frac{d}{w} \tag{12}$$

we can substitute the above relationships to get

$$R_S \alpha \frac{L}{\sqrt{L}} = \sqrt{L} \tag{13}$$

That is, resolution is proportional to the square root of the column length.

This relationship can also be shown graphically (Figure 4.1), where the distance d and the peak width 4σ must be in the same units. At some

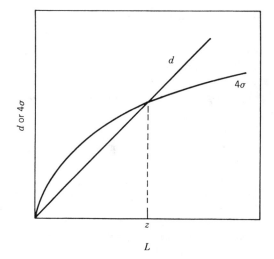

Figure 4.1. The achievement of separation. Adapted from J. C. Giddings, *Dynamics of Chromatography*, Part 1, Marcel Dekker, New York, 1965, p. 33. Courtesy of Marcel Dekker.

column length indicated by the dashed line, d will exceed 4σ, the peak width, and the two analytes will be separated according to the definition of resolution [Eq. (12)]. Our conclusion is that, as long as two analytes have some difference in their partition coefficients, it must be possible to make a column long enough to separate them. In practice, of course, this is not usually the easiest way to get a separation, and some extremely long columns would be impossible to work with.

THE OPTIMIZATION OF SEPARATIONS

What are the best ways to optimize separations? One way to answer this question is to consider a popular form of the resolution equation that is originally attributed to Purnell.[1]

$$R_s = \frac{1}{4}\sqrt{n}\left[\frac{(\alpha - 1)}{\alpha}\right]\left(\frac{k_B}{1 + k_B}\right) \tag{14}$$

The three variables are n (column efficiency), α (column selectivity), and k_B, which is the partition ratio for the second of the two analytes being separated. (Several similar equations are also in common use, and the relationships between them have been discussed by Said.[2]) Since good resolution is indicated by a large value for R_s, the three terms, $\sqrt{n}$, $(\alpha - 1/\alpha)$, and $(k/1 + k)$ should be maximized. However, time is a fourth variable that also needs to be considered. That discussion will follow our consideration of Eq. (14).

The number of theoretical plates can easily be increased by increasing the length of the column; we have just seen how that concept works. However, the resolution is increased only by the square root of the increase in length at the expense of a linear increase in time. For this reason alone, longer columns are not usually the preferred method for improving resolution.

Alternatively, n can be increased by preparing a better column according to the rate equation. The most important parameters are stationary film thickness and particle diameter, both of which should be kept small. Further discussion can be found in Chapter 2.

The next term, containing α, the selectivity or separation factor, will be maximized if α is maximized. In effect this means that the separation system, the stationary and mobile phases, should be chosen to show the greatest selectivity between analytes A and B; our knowledge of intermolecular forces can help us in selecting the best system. Typical values of α range from just over 1 to nearly 2 (which would be an easy separation). Thus the values of the term $(\alpha - 1)/\alpha$ could vary from 0.001 up to 1, a range of 10^3, indicating that this term is the most powerful of the three.

Finally, the partition ratio k should also be maximized in order to maximize the last term, $k_B/(1 + k_B)$. Since the two analytes are next to each other, k_B is not much different from k_A, and it is sufficiently accurate to merely indicate that k (either one) should increase. However, a few calculations will show that little improvement in resolution is gained as k is increased above about 10. For example, if k is increased from 1 to 50, the increase in this term is from 0.5 to about 1. The improvement in resolution is not nearly as significant as that for α, and any increase in k denotes a proportionate increase in time. The optimum value for k should consider not only the resolution but also the time.

We can get a time-dependent equation by substituting Eq. (15) into Eq. (14)

$$t_R = \frac{n(k + 1)H}{\bar{u}} \tag{15}$$

to give

$$t_R = 16R_s^2 \left(\frac{\alpha}{\alpha - 1}\right)^2 \frac{(k + 1)^3}{(k)^2} \frac{H}{\bar{u}} \tag{16}$$

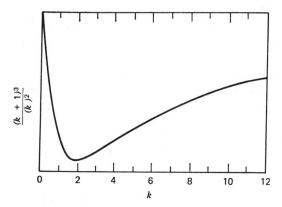

Figure 4.2. Time-optimized partition ratio.

To find the minimum time for a given resolution, the k term is plotted versus k as shown in Figure 4.2. The minimum is a k value of around 2, but most chromatographers have found that k values in the range of 1.5 to 5 represent an optimal range. The only disadvantage in higher k values is the additional time required for analysis, and since GC is very fast, many GC runs are made at higher k values. In practice, the k value is adjusted by varying the temperature or by varying the phase volume ratio (β) by changing the amount of stationary phase. To increase k at constant temperature, β must be decreased.

Also, in LC, small changes in the composition of the mobile phase can be used to change k. Several papers by Snyder[3] and one by Grushka[4] discuss the best methods for choosing experimental conditions for LC.

Equation (14) can be rearranged and used to calculate the number of plates required for a given separation:

$$n_{required} = 16\, R_s^2 \left[\frac{\alpha}{(\alpha - 1)} \right]^2 \left[\frac{(k + 1)}{k} \right]^2 \qquad (17)$$

Table 1 gives a number of typical values of n calculated with Eq. (17). It can be seen how quickly n decreases as k is increased up to about 5, and how a small change in α from 1.05 to 1.10 decreases the plates needed by a factor of about 4.

One final variation of Eq. (14) can be derived if it is assumed that all peaks in a given separation have the same plate number:

$$R_s = \frac{1}{2}\, \sqrt{n}\ \frac{(k_B - k_A)}{(2 + k_A + k_B)} \qquad (18)$$

TABLE 1 Number of Theoretical Plates Required to Achieve a Given Resolution

Capacity Factor, k_B	$R_s = 1.5$		$R_s = 1.0$	
	$\alpha = 1.05$	$\alpha = 1.10$	$\alpha = 1.05$	$\alpha = 1.10$
0.1	1,921,000	527,080	853,780	234,260
0.2	571,540	156,820	254,020	69,700
0.5	142,890	39,200	63,500	17,420
1.0	63,500	17,420	28,220	7,740
1.5	44,100	12,100	19,600	5,380
2.0	35,720	9,800	15,880	4,360
5.0	22,860	6,273	10,160	2,790
10.0	19,210	5,270	8,540	2,340
20.0	17,500	4,800	7,780	2,130
∞	15,880	4,360	7,060	1,940

This equation is used in Chapter 9 in the discussion regarding the optimization of the mobile phase in LC.

REFERENCES

1. J. H. Purnell, *J. Chem. Soc.*, **1960**, 1268.
2. A. S. Said, *HRC CC, J. High Resolut. Chromatogr. Chromatogr. Commun.* **1979**, *2*, 193–194.
3. L. R. Snyder, *J. Chromatogr. Sci.* **1972**, *10*, 200, 369.
4. E. Grushka, *J. Chromatogr. Sci.* **1972**, *10*, 616.

COMPARISONS BETWEEN CHROMATOGRAPHIC MODES 5

Before proceeding with a discussion of the specifics of individual chromatographic techniques, it will be helpful to take a look at some of the differences between various modes. The emphasis in this monograph has been on similarities between modes, but there are significant differences that should be examined.

GAS CHROMATOGRAPHY COMPARED TO LIQUID CHROMATOGRAPHY

A comparison of the two major types of chromatography, GC and LC, reveals that the major differences can be attributed to the differences in properties between gases and liquids. The two most important parameters in GC are (1) the nature of the stationary phase and (2) temperature. In LC they are (1) the nature of the stationary phase and (2) the nature of the mobile phase. In general, GC can be used only with volatile samples (those that boil below 500°C or have a vapor pressure of several kilopascals) that are also thermally stable, and this is not a limitation imposed on LC. For further comparison, we can examine the differences between gases and liquids.

Gases Compared to Liquids

Giddings has written extensively on this subject and has provided a thorough discussion in his book.[1] First of all, what he refers to as the "density of intermolecular attraction" is much lower for gases than for liquids— by about 10^4. This means that the mobile phase in GC does not interact with analytes and cause them to desorb from the stationary phase. It simply carries them down the column when they are in the vapor state. This lack of intermolecular attraction is normally desired in GC where the gases used (He and N_2) are chosen for their inertness. In LC, the opposite is true, and the mobile phase liquids compete actively with the stationary phase to attract the analytes. In order to get this enhanced selectivity in GC, some workers have tried adding condensable gases to their mobile phases, but the main effect has been a partial deactivation of the stationary phase.[2]

Other comparisons have more to do with operating conditions than with fundamental mechanisms of interaction. For example:

1. The diffusivity of liquids is less than gases by a factor of about 10^{-5}. This was noted in Chapter 2, where some typical diffusion coefficients are given. The slower diffusion in liquids results in slower speeds of analysis in LC and in a significant contribution by the mass transfer term C_M in causing zone broadening.

2. The viscosity of liquids is greater by a factor of about 10^2. One consequence is the higher operating pressures for LC.

3. The surface tension of liquids is greater by about 10^4. (Gases can be considered to have no surface tension.) The surface tension of liquids makes possible ascending TLC and PC due to the driving force of capillarity.

4. The density of liquids is greater by about 10^3.

5. Liquids are virtually incompressible, and gases are readily compressed. This has a major impact on GC because the mobile phase carrier gas is compressed at the head of the column, resulting in a variable flow rate through it. We need to take a closer look at this phenomenon and see what the consequences are for GC operating parameters.

Gas Compressibility in GC. Figure 5.1 shows the effect of the pressure gradient in the column on carrier gas flow. The carrier gas is under higher pressure at the column inlet (P_i) than at the column outlet (P_o), which is usually at atmospheric pressure. The various curves show that the changes in flow rate are greater at the higher ratios of inlet-to-outlet pressure (P_i/P_o). Thus the least change in flow is encountered when the column pressure drop is least, as is the case with open tubular columns.

Usually the flow rate is measured at the exit at P_o where the rate is at its maximum. To get the average flow in the column, this maximum value must be multiplied by a correction factor called the *compressibility, j.*

$$j = \frac{3}{2}\left[\frac{(P_i/P_o)^2 - 1}{(P_i/P_o)^3 - 1}\right] \tag{1}$$

Table 1 lists some values for j, and it can be seen that this is a significant factor. If the symbol F_c is used to denote the exit flow at column tem-

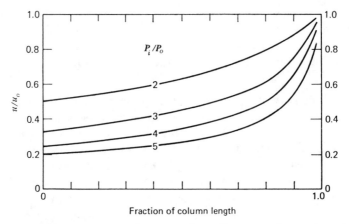

Figure 5.1. Effect of pressure drop on column flow at various points along the column length.

perature, then the average flow is

$$\overline{F}_c = jF_c \tag{2}$$

Similarly,

$$\text{average pressure} = \overline{P} = \frac{P_o}{j} \tag{3}$$

$$\text{average linear velocity} = \overline{u} = ju \tag{4}$$

$$\text{corrected retention volume} = V_R^0 = jV_R \tag{5}$$

$$= jF_c t_R \tag{6}$$

As indicated in Eq. (5), when the compressibility factor is applied to the

TABLE 1   Some Pressure Correction Factors (j)

P_i/P_o	j	P_i/P_o	j
1.1	0.95	1.8	0.70
1.2	0.91	1.9	0.67
1.3	0.86	2.0	0.64
1.4	0.83	2.2	0.60
1.5	0.79	2.5	0.54
1.6	0.76	3.0	0.46
1.7	0.72	4.0	0.36

retention volume, the adjective *corrected* is used and a superscript zero is added to the symbol. The same convention applies to all gas volumes, such as the mobile phase volume, which can be denoted by a subscript $M (V_M^0 = jV_M)$ or a subscript G for gas $(V_G^0 = jV_G)$.

The flow at the outlet, F_o, is usually measured at ambient temperature with a soap film flow meter. In that case, several corrections must be applied to get the desired flow F_c.

$$F_c = F_o \frac{T_c}{T_a} \left(\frac{P - p_w}{P} \right) \tag{7}$$

T_a and T_c are the ambient temperature and the column temperature, respectively, in degrees Kelvin; P is the atmospheric pressure and p_w is the vapor pressure of water at T_a—both pressures are in the same units. In some cases in this monograph the symbol F is used with no special designations to refer to a nonspecific flow rate.

When a GC retention volume is both *corrected* (for average flow) and *adjusted* (by subtracting out V_G), a new name and symbol are used.

$$jV_R' = V_R^0 - V_G^0 = V_N \tag{8}$$

V_N is called the *net* retention volume. Since an analogous situation does not exist in LC, it is possible that some confusion may arise over the use of different terms. Thus, we originally used the equation

$$V_R' = KV_s \tag{9}$$

in our definitions in Chapter 1, and that was satisfactory for LC, but we should use

$$V_N = KV_s \tag{10}$$

for GC. Often Eq. (10) and the term *net retention volume* are also used for LC, just to reduce confusion and make one equation serve both. No confusion should arise if the equations are read in context.

Permeability. One final difference between gases and liquids results in a difference in permeability between columns in GC and LC. As already noted, LC columns require much higher inlet pressures than corresponding GC columns, so this parameter is of primary interest in LC. Permeability, κ, is defined as

$$\kappa = \frac{\eta L \bar{u}}{\Delta P} \qquad (11)$$

where η is the viscosity, L is the column length, $\bar{u}$ is the average mobile phase velocity, and ΔP is the pressure drop across the column. The average mobile phase velocity can be measured as L/t_M, and substituting this value into Eq. (11) gives an equation from which permeability can be experimentally measured:

$$\kappa = \frac{\eta L^2}{\Delta P t_M} \qquad (12)$$

However, this equation does not include one of the most important LC variables, the particle diameter d_p. Consequently, another parameter has been defined[3] that does include it. It is called the *flow resistance parameter*, Φ:

$$\Phi = \frac{d_p^2}{\kappa} = \left(\frac{d_p}{L}\right)^2 \frac{\Delta P t_M}{\eta} \qquad (13)$$

In calculating Φ is is important to use the proper units; those most commonly used by chromatographers are: ΔP in bar (10^5 Pa), t_M in seconds, d_p in μm, η in centipoise (mN s/m^2), and L in centimeters. The most permeable columns are the most desirable, and they will have the smallest values of Φ. Typically LC columns will have values ranging from about 500 for spherical microparticles to 1000 for irregular microparticles.

Bristow and Knox[3] also defined some other parameters for characterizing LC columns and discussed in detail the necessity to standardize test conditions for LC. Further discussion of this topic can be found at the end of Chapter 9.

Supercritical Fluids

There is a region in a gas–liquid phase diagram where a substance can exist with properties characteristic of both the gas and liquid states. This state is referred to as a *fluid*, or more precisely, a *supercritical fluid*, since it exists in the region beyond the critical points for the substance. Visualize a transparent container half filled with liquid and sealed. As its temperature is increased its pressure will increase until the critical point is reached. At that point, the meniscus between the two phases disappears and the material in the tube is in the supercritical phase. A typical phase diagram is shown in Figure 5.2 for carbon dioxide, indicating the critical point.

In the mid-1960s, as high pressures became common in LC, it was only natural to see if the supercritical region had any special chromatographic properties. It did, and the field of supercritical fluid chromatography (SFC) was born. More details can be found in Chapter 11.

Efficiency and Speed

Is either GC or LC inherently faster or more efficient? Giddings[1] attempted to answer that question by deriving the relationship

$$\frac{(n_{\lim})_{LC}}{(n_{\lim})_{GC}} = \frac{3}{2} \frac{\eta_G D_G}{\eta_L D_L \Delta P} \tag{14}$$

which compares the limit on the plate numbers for the two techniques. Substituting the appropriate viscosity and diffusivity ratios (discussed earlier) and allowing for the compressibility of gases, he found that

$$\frac{(n_{\lim})_{LC}}{(n_{\lim})_{GC}} \approx \frac{10^3}{\Delta P} \tag{15}$$

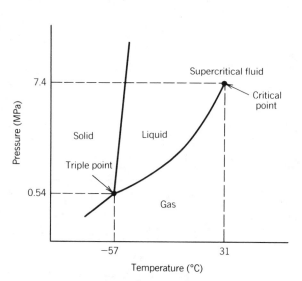

Figure 5.2. Phase diagram of carbon dioxide.

where ΔP is in atmospheres. In effect, if GC and LC were run at the same pressures, LC would be more efficient by a factor of 10^3, or if GC were run at a pressure 10^3 times as large as LC, it would be as efficient as LC. The problem with the first conclusion is that the time required to achieve the large plate numbers in LC would be very long, as shown in Figure 5.3. (Note that the figure is a log–log plot and the time axis gets large very quickly!) The problem with the second conclusion is that it has been found difficult to run GCs at the high pressures required by the equation.

Probably it is safe to say that the theory has not provided us with much guidance, and the choice between GC and LC is seldom made on the basis of large differences in efficiencies. In most cases, a sample that is amenable to GC—that is, one that is volatile and stable—can be run faster and more easily by GC than by LC despite the prediction that LC should be more efficient.

Reduced Parameters

Some comparisons of the efficiencies of chromatographic separations are facilitated by defining so-called *reduced parameters*, which correct for

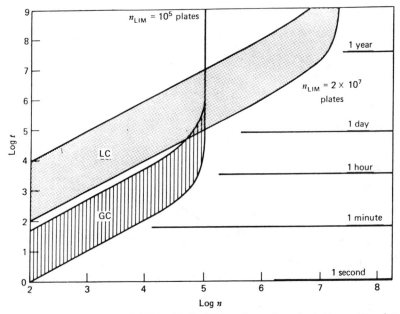

Figure 5.3. Comparison of GC and LC efficiencies and analysis times. Reprinted with permission from J. C. Giddings, *Anal. Chem.* **1965**, *37*, 60. Copyright 1965, American Chemical Society.

the differences between gases and liquids. For example, if one wants to describe the rate at which analytes move in the mobile phase, one needs to compare the rates of diffusion of analytes in gases and liquids, and contrast them with the flow rate of the mobile phase; analytes can reach the sites on the stationary phase only by diffusion or by being swept along by the mobile phase. As we have seen, the relative rates of diffusion between gases and liquids differ by a factor of 10^4 or 10^5, and the flow rates in LC are usually much slower than in GC. Therefore, a *reduced* parameter that eliminates these differences would be very helpful in comparing the performance of an LC column with that of a GC column.

The two parameters in the rate equation that have been redefined are the column diffusivity H and the linear velocity u. The reduced parameters, which are dimensionless, are

$$h = \frac{H}{d_p} \tag{16}$$

and

$$v = \frac{ud_p}{D_M} = \frac{Ld_p}{t_M D_M} \tag{17}$$

The rate equation given in these reduced parameters is

$$h = \frac{B}{v} + (C_s + C_M)v \tag{18}$$

Figure 5.4 shows plots of the reduced rate equation (on a log–log basis) taken from Knox and Saleem.[4] GC is compared with LC for two stationary phases and several nonretained analytes. While all the points do not fall on one smooth curve, the plot clearly shows the similarities between GC and LC plots when reduced parameters are used.

The reduced parameters are also helpful in evaluating column performance. The best columns have a reduced plate height of 2 to 5—a number that can be thought of as representing the number of particles between sorptions—and 2 is a practical minimum. The reduced velocity represents the ratio between the flow velocity and the diffusion rate over one particle diameter; typical values should be in the range of 3 to 20.

Bristow and Knox[3] have also determined a reduced rate equation for LC by fitting their data to the equation

$$h = \frac{B}{v} + Av^{0.33} + Cv \tag{19}$$

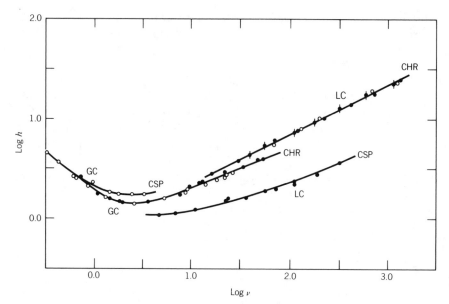

Figure 5.4. Comparison of plots of reduced Rate Equations for GC (on left) and LC (on right). CHR = Chromosorb G; CSP = duPont CSP beads. Reproduced from the *Journal of Chromatographic Science* by permission of Preston Publications, Inc.

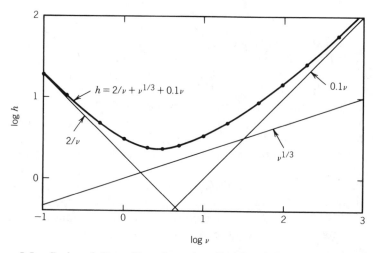

Figure 5.5. Reduced Rate Equation plot. Reprinted from reference 3 with permission.

As shown in Figure 5.5, they find that a good LC column fits Eq. (20)

$$h = \frac{2}{\nu} + \nu^{0.33} + 0.1\nu \tag{20}$$

and it has become an accepted model for LC columns.

COLUMN AND PLANAR CONFIGURATIONS

In GC, the stationary phase is always packed in a column in order to contain the mobile phase, a gas. A similar configuration is common in LC, but since the mobile phase is a liquid, it is also possible to use another configuration—one in which the stationary phase is spread on a flat, plane surface. This configuration is called either *paper chromatography (PC)* or *thin layer chromatography (TLC)*, depending upon the stationary phase. In TLC, the stationary phase is coated on a supporting planar surface, which can be glass, plastic, or metal. Many of the stationary phases used in columnar LC can also be used in TLC, so it is interesting to compare these two techniques. As a background for that comparison, we first need to define a few other terms and symbols.

Peaks Compared to Bands

Up to this point it may have appeared that the terms *band, peak*, and *zone* have been used synonymously, but let us take a closer look. As commonly used, all three terms describe the distribution of analyte molecules in space (a concentration profile), but *band* represents this distribution while the analyte is still *in* (or *on*) the system, while *peak* refers to a distribution of analyte that has eluted from the system. The term *zone* is more general and includes both bands and peaks; it is used in those cases when we do not wish to be more specific. In the context of our current discussion, this means that the zones in TLC and PC are called bands and those in column LC and in GC are called peaks.

An attempt to visualize the difference between bands and peaks is given in Figure 5.6, which shows two analytes, A and B. The partition coefficient for A, K_A, is greater than the partition coefficient for B, K_B. Consequently, B migrates faster than A, and as shown in Figure 5.6a, B has moved farther down the bed than A; in Figure 5.6b, B is shown eluting from the bed before A. As noted in the figure, chromatograms of the two situations show the zones in reverse order and width.

Thus, zone broadening does not depend on the length L of the bed, as

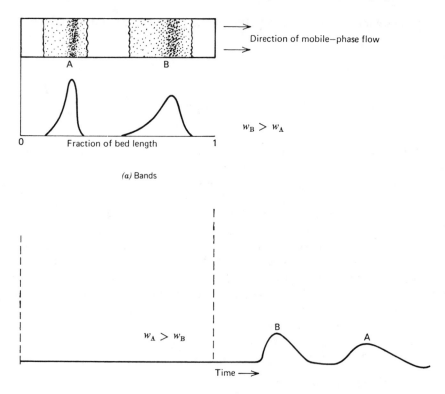

Figure 5.6.   Illustration of the difference in presentation of bands and peaks; K_A > K_B.

suggested by the definition of column dispersivity introduced in Chapter 2:

$$H = \frac{\sigma^2}{L} \qquad (21)$$

Rather, *band* dispersivity is a function of the distance migrated, L_z, in the case of TLC,

$$\sigma = \sqrt{HL_z} \qquad (22)$$

and *peak* dispersivity is a function of the time spent in the system, the retention time t_R,

$$\sigma = \sqrt{Ht_R} \qquad (23)$$

Also, for the bands in Figure 5.6, $\sigma_B > \sigma_A$, while for peaks the opposite is true. While most of this text deals with column techniques and peaks, the distinction between peaks and bands needs to be maintained for clarity.

Retention Ratio and Retardation Factor

There is another difference between column techniques and planar techniques in LC. It is the way that the retention ratio is expressed. In Chapter 1, the retention ratio R_R was defined for column techniques as

$$R_R = \frac{t_M}{t_R} = \frac{v}{u} \tag{24}$$

and the definition is illustrated in Figure 5.7b. In TLC and PC, the analogous ratio is usually called the *retardation factor*, and the symbol used is R_f.

$$R_f = \frac{\text{distance migrated by an analyte}}{\text{distance migrated by the solvent front}} \tag{25}$$

It is illustrated in Figure 5.7a. Clearly, both R_R and R_f can be easily calculated from their respective chromatograms, making them very useful measures for describing chromatographic results. Recall also that $R_R = 1/(1 + k)$ and that k is proportional to the partition coefficient K, the basic thermodynamic variable in chromatographic theory.

Chapter 10 contains more information about TLC and PC, but it should

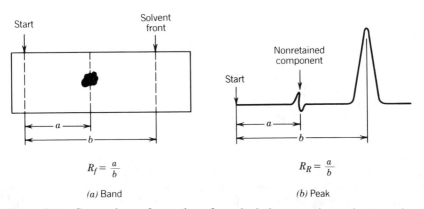

Figure 5.7. Comparison of equations for calculating retention ratio R_R and retardation factor R_f. Reprinted courtesy of Gow-Mac Instrument Co.

be noted here that the reason the planar techniques have a retention ratio different from the column methods is that the flow cannot be controlled, and consequently it is not constant in the planar methods. As a result, we are forced to measure "distances" on the planar surfaces rather than retention times or volumes, and the calculated R_R and R_f values will not be exactly the same.

To see how the two methods compare, some data published by Majors[5] are given in Figure 5.8, which shows the TLC and column separations of six azo dyes, listed in Table 2. The similarity between the two separations

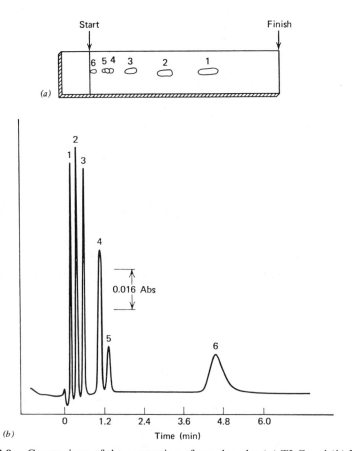

Figure 5.8. Comparison of the separation of azo dyes by (a) TLC and (b) LSC. Conditions for TLC: silica gel F-254; solvent 10% CH_2Cl_2 in hexane. Conditions for LSC: 15 cm × 2.4 mm column of MicroPak SI-10; solvent same as TLC; flow rate, 132 mL/hr; pressure, 350 psi. See Table 2 for identification of dyes. Reprinted with permission from R. E. Majors, *Anal. Chem.* **1973,** *45,* 755. Copyright 1973, American Chemical Society.

TABLE 2   Dyes Separated in Figure 5.8

Number	Structure	TLC R_f	LSC $R_R{}^a$
1	⟨benzene⟩—N=N—⟨benzene⟩	0.69	0.55
2	⟨benzene(Br)⟩—N=N—⟨benzene⟩—NEt$_2$	0.39	0.32
3	⟨benzene⟩—N=N—⟨benzene⟩—NEt$_2$	0.21	0.21
4	⟨benzene(NO$_2$)⟩—N=N—⟨benzene⟩—NEt$_2$	0.098	0.12
5	O$_2$N—⟨benzene⟩—N=N—⟨benzene⟩—NEt$_2$	0.064	0.097
6	⟨benzene⟩—N=N—⟨benzene⟩—NH$_2$	0.015	0.029

^{a} Estimated from Figure 5.8b.

is obvious from the figure, but it is better evaluated by comparing the R_R and R_f values in Table 2. The R_f values are taken from the paper, but the R_R values had to be estimated, since no t_M was given. The two columns of numbers clearly show the good agreement between these two parameters and suggest that TLC data should be useful in designing column LC methods. It is instructive to note again that the peaks in Figure 5.8 are in reverse order and reverse width compared to the bands.

Also, there are some operational differences that cause differences between the retention ratio and the retardation factor. Even if the exact same stationary phase is coated on a TLC plate and packed in a column, the TLC material usually contains an additional binder to hold the stationary phase on the plate. This binder will most likely alter slightly the characteristics of the stationary phase and result in differences between the R_R and R_f values. And, finally, in normal TLC procedures the stationary bed is usually dry when the chromatographic elution process is begun, whereas it is usually wet with mobile phase in the column processes. This difference may also contribute to differences between retention parameters.

SPECIAL TECHNIQUES

The column and planar configurations just discussed are the most common modes of LC but not the only ones. At least four different methods have been proposed for performing LC without a solid support, and two have survived as viable methods. The term *countercurrent chromatography* is applied to these techniques, and a review of their development has been written by Ito and Conway.[6] Countercurrent chromatography will not be discussed further here, but it should be noted that it has one main advantage: it is free from undesirable adsorption and catalytic effects that sometimes result from the use of solid supports.

Somewhat similar is a continuous GC method that does not use a packed bed.[7] It consists of a pair of parallel disks on which the stationary phase is coated. The sample enters at the center, and both disks are rotated.

One final variation is the use of centrifugal force to move the mobile phase over the stationary phase.[8] This technique has been called *centrichromatography*, and a commercial apparatus is available.

REFERENCES

1. J. C. Giddings, *Dynamics of Chromatography*, Part 1, Dekker, New York, 1965, pp. 293–301.
2. J. F. Parcher, *J. Chromatogr. Sci.* **1983**, *21*, 346.
3. P. A. Bristow and J. H. Knox, *Chromatographia* **1977**, *10*, 279.
4. J. H. Knox and M. Saleem, *J. Chromatogr. Sci.* **1969**, *7*, 745.
5. R. E. Majors, *Anal. Chem.* **1973**, *45*, 755.
6. Y. Ito and W. D. Conway, *Anal. Chem.* **1984**, *56*, 534A; Y. Ito, *Adv. Chromatogr. N. Y.* **1984**, *24*, 181.
7. M. V. Sussman and C. C. Huang, *Science* **1967**, *156*, 974.
8. F. W. Karasek and P. W. Rasmussen, *Anal. Chem.* **1972**, *44*, 1488.

QUALITATIVE ANALYSIS 6

The chromatographic parameter that can be used for qualitative analysis is the retention time or retention volume. For a defined system with invariant conditions, a given analyte will have a constant retention volume, and an unknown can be identified by comparison with retention volumes of standards. Figure 6.1 contains two chromatograms run under the same conditions, one with an unknown (Figure 6.1a) and one with five known alcohols. The five peaks in the first chromatogram are probably the five alcohols with the same respective retention times. However, too many variables must be closely controlled to make this relationship useful except when samples are run in close time proximity on the same system; slight changes in columns, temperatures, flow rates and other variables make it useless otherwise.

Another drawback is the fact that there are over 60,000 chemical compounds in commerical use, and each one cannot have a distinctive retention volume on a given system. If a compound or a mixture is truly unknown, its chromatogram will not provide characteristic retention volumes on the basis of which identifications can be made. In short, chromatography can be used for qualitative analysis for a limited set of chemicals, but its main use is not for screening unknowns. In fact, one of the best methods of qualitative analysis is the combination of chromatography (an excellent separation method) with mass spectrometry (an excellent identification method). This topic is included in Chapter 11.

RETENTION PARAMETERS

Before completely abandoning retention volume as a characteristic parameter for qualitative analysis, we should review its various forms. We have already defined the following terms:

$$\text{retention volume} = V_R = t_R \times F \tag{1}$$

$$\text{adjusted retention volume} = V'_R = V_R - V_M \tag{2}$$

$$\text{corrected retention volume} = V^0_R = jV_R \tag{3}$$

$$\text{net retention volume} = V_N = jV'_R \tag{4}$$

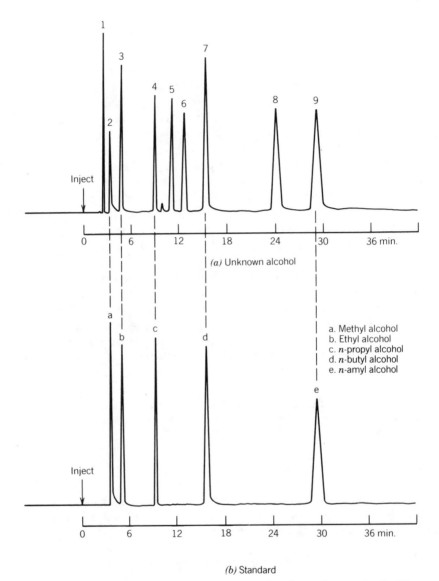

(a) Unknown alcohol

a. Methyl alcohol
b. Ethyl alcohol
c. *n*-propyl alcohol
d. *n*-butyl alcohol
e. *n*-amyl alcohol

(b) Standard

Figure 6.1. Identification of unknowns by retention time using standards. Figure courtesy of McNair and Bonnelli, *Basic Gas Chromatography*, Varian, 1968.

Net retention volume is the one term that would be expected to be the most reproducible, but it is not good enough to be tabulated. In the early days of GC, the *specific* retention volume V_g was also defined for use in qualitative analysis, but it was not found to be useful.

Relative Retention Parameters

It is likely that a relative retention volume that is ratioed to a standard would be more reproducible. Such has been found to be true; α is a relative retention ratio; it was defined in Chapter 1, where it was called the *column selectivity* or a *separation factor*:

$$\alpha = \frac{K_B}{K_A} = \frac{(t'_R)_B}{(t'_R)_A} \tag{5}$$

The main problem with α is that there is no single standard to which data have been ratioed, and consequently there are no tabulations of relative retention data in the literature. Kovats[1] suggested that a series of standards be used, and he proposed the *n*-paraffins. Before discussing his proposal further, we need to examine the relationship between retention volume and the members of a homologous series such as the paraffins.

An isothermal GC run of paraffins on a given column will result in a chromatogram like that shown in Figure 6.2, where the retention volumes increase logarithmically with carbon number. This is the expected behavior since we know that

$$\log K = -\frac{\Delta \mathscr{H}}{2.3 \, RT} + \text{constant} \tag{6}$$

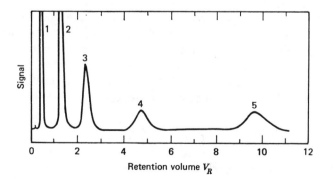

Figure 6.2. Typical chromatogram for a homologous series of compounds.

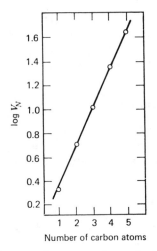

Figure 6.3. Plot of carbon number versus log net retention volume for a homologous series.

where $\Delta \mathcal{H}$ is the enthalpy of vaporization, which can be estimated by Trouton's rule as

$$\Delta \mathcal{H} = 21 T_{\text{boil}} \tag{7}$$

Since the boiling points T_{boil} of the members of a homologous series increase in a regular fashion, so should $\Delta \mathcal{H}$; K should increase logarithmically, as should V_N since $V_N = K V_s$. When log V_N is plotted versus the carbon number, as shown in Figure 6.3, a straight line results for most homologous series.*

Kovats Index. Kovats[1] defined a retention index I in which the n-paraffins are assigned reference values 100 times their respective carbon numbers. Thus hexane has an index of 600, heptane 700, and so on. When they are run isothermally on a specified column, and the logs of their net retention volumes are plotted versus carbon number, a straight line results (Figure 6.4), as expected from the homologous series relationship just described. If any other compound is run under the same conditions, its index can be read from the graph using the log of its net retention volume. Alternatively, the index can be calculated as follows:

$$I = 100 \left[\frac{(\log V_N)_u - (\log V_N)_x}{(\log V_N)_{x+1} - (\log V_N)_x} \right] + 100x \tag{8}$$

---

* Remember that V_N is proportional to t_R', which can be used instead if flow rate and pressure drop are constant.

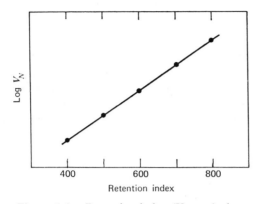

Figure 6.4. Retention index (Kovats) plot.

where u stands for the unknown, x for the paraffin with x carbons and eluting just before the unknown, and $x + 1$ for the paraffin with $x + 1$ carbons and eluting just after the unknown.

The retention index has become the standard method for reporting GC data. By definition, the members of *any* homologous series should differ from each other by 100 units just as the standards do. This relationship is not always exact,[2] and the index is somewhat temperature dependent.[3] Nevertheless, it is very popular, and McReynolds[4] has published a self-consistent book of indices for 350 compounds on 77 stationary phases and at 2 temperatures. Other homologous series have also been used as standards in specific industries.

In Chapter 8 we will see that programmed temperature GC results in a regular, linear relationship between retention time and carbon number. Under those circumstances logs should not be used in Eq. (8) and in the retention index plot. The increase in temperature decreases the partition coefficients and effectively removes the logarithmic dependence of I.

For LC a similar relationship should apply if the retention mechanism shows the expected theoretical dependence on carbon number. The situation is more complex since the partition coefficient is a function of many intermolecular forces. Several papers have been published showing a homologous series retention like that described for GC.[5] In principle then, the retention index concept should also apply in those cases. However, little interest has been shown in developing an index for LC, probably because the paraffins are not usually run by LC and the modes of analysis by LC are much more variable and complex, so that the data are not as widely usable.

Two Column Plots

Qualitative analysis is enhanced if data are acquired on more than one system. For example, in GC it is fairly common and easy to run a sample on each of two columns that are chosen to be widely different in their polarities. The results can be plotted as net retention volumes or as Kovats index values on either linear or log scales as shown in Figure 6.5. In either case, straight lines result for homologous series, thus aiding qualitative identifications. The principle is simple: the more data, the more reliable the analysis.

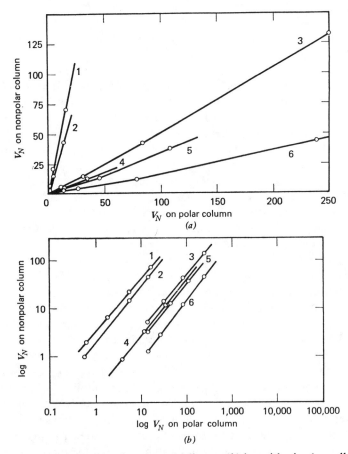

Figure 6.5. Homologous series plots: (*a*) linear; (*b*) logarithmic. 1 = alkanes, 2 = cycloalkanes, 3 = esters, 4 = aldehydes, 5 = ketones, 6 = alcohols. Reprinted with permission from J. S. Lewis, H. W. Patton, and W. I. Kaye, *Anal. Chem.* **1956**, *28*, 1370. Copyright 1956, American Chemical Society.

Two column plots are more effective if the two stationary phases are very different from each other. If they are of *opposite* polarity, then the elution order of the analytes in a given sample should be very different (opposite) on the two phases. For example, in the GC separation of a mixture of *n*-heptane, tetrahydrofuran, 2-butanone, and *n*-propanol, the elution order on a polar phase like Carbowax is in the order just listed; on a nonpolar phase like SE-30, the order is the exact opposite! Neither separation follows the boiling point order.

In LC, similar reversals can be expected when the stationary and mobile phases are reversed in polarity. In fact, the effect is so marked that LC methods are classified as *normal* when the stationary phase is polar and the mobile phase is nonpolar and as *reversed* when the stationary phase is nonpolar and the mobile phase is polar. The elution order of analytes should be opposite (reversed) in one mode compared to the other. This effect is *expected* in chromatography, but it is easy to forget it and assume that the elution order is always the same regardless of the column.

OTHER METHODS OF QUALITATIVE ANALYSIS

Virtually every technique imaginable has been examined in an attempt to improve chromatography's ability to perform qualitative analyses.[6] Many are specific for only one sample type or only one chromatographic procedure, but some typical examples will be discussed to indicate the range of possibilities. They have been divided into chemical methods and instrumental (detector) methods.

Chemical Methods

The chemical methods can be divided into those that are applied to the sample before it is analyzed and those that are applied after the analytes have been chromatographed. In the former category, the so-called precolumn reactions are many derivatizations. As discussed later, derivatives are usually made to facilitate their being chromatographed (volatility, detectability, and so on), but in some cases the derivatives may chromatograph differently and provide qualitative information. One example is the reactor of Beroza,[7] which was used with GC to remove functional groups prior to analysis. Similarly, some reactions can be carried out *in situ* in the injection port of a GC because of its high temperature, which produces increased reaction rates between the sample and added reagents.

Pyrolysis is another common precolumn treatment that has been used in GC.[8] Nonvolatile samples are pyrolyzed very rapidly at high temperatures, and the degradation products are chromatographed. Polymers can be distinguished by the pattern of peaks obtained, without actual identification of the individual peaks.

Chemical reactions are used to identify analytes after they have been separated. In TLC, it is common to spray the plate with a chemical that will selectively react with certain analytes. In GC, the column effluent can be bubbled into a solution containing a derivatizing reagent that will form a color or a precipitate[9] similar to the well-known spot tests. In LC, the column effluent can be mixed with a second reagent stream to produce a postcolumn reaction for purposes of identification or simply for detection.[10]

Instrumental Methods

The instrumental methods are focused on the detector or an auxiliary instrument used as a supplementary detector.

Single Selective Detectors. In both GC and LC there are detectors that show a selectivity for particular groups of compounds or functional groups. Conventional examples will be given later in the individual chapters, but one unusual, specific detector is the moth (alive) used with a GC to detect the presence of sex pheromones in a column effluent.[11] The physical response of the moth clearly indicates which peak represents its sex hormone. Humans also sniff column effluents to identify particular odors in the flavor and fragrance industry.

Dual Detectors. Most dual detectors are run in parallel, the column effluent being split and run through both of them simultaneously. In GC the technique is known as *dual channel GC*; usually, one of the detectors chosen is universal and the other is highly selective. Figure 6.6 shows the analysis of a commerical gasoline sample with dual detection by flame ionization (FID) and electron capture (ECD). The FID detects all the hydrocarbons, but the ECD is selective for the alkyl lead additives in gasoline and permits their detection without interference from the hydrocarbons.

Another example[12] is the separation of atmospheric hydrocarbons on an FID and a photoionization detector (PID), which is more sensitive for unsaturated hydrocarbons (Figure 6.7). The identities of the peaks are given in Table 1 along with the PID/FID response ratios. It can be seen that the ratios can be used as additional information in assigning peak identities. In fact, Figure 6.8 shows that the hydrocarbon type (saturates, olefins, or aromatics) can be assigned from the detector ratio in many cases.

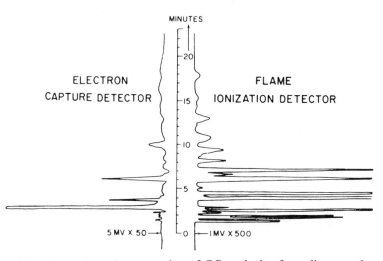

Figure 6.6. Dual channel presentation of GC analysis of gasoline sample on a packed DC-200 column. Courtesy of Perkin-Elmer Corp.

A similar arrangement in LC is the simultaneous, parallel detection of column eluents at two wavelengths in a UV detector. In this case, the two signals are usually ratioed against each other, giving rise to a chromatogram, or *ratiogram*, that produces squared-off peaks for individual components. Figure 6.9 shows four chromatograms and four ratiograms of an unknown mixture run with four different mobile phases.[13] The ratiogram was helpful in determining how many analytes were in the sample, four or five. Note that a ratio near 1 does not produce a peak (e.g., peak 1 in run *b*), and that a well-resolved peak gives a squared-off peak (peak 2 in run *b*). The third peak in ratiogram *b* is not square, indicating a possible overlapping impurity; the presence of this fifth component is confirmed in chromatogram *d*.

The availability of multiple wavelength UV detectors (see photodiode array detectors in Chapter 9) makes possible analyte recognition by increasing the number of independent informational degrees of freedom. A computer is needed to handle the data. Some recent examples have been given in LC[14] and SFC.[15] Mathematical deconvolution of peaks containing up to three components has been reported[16] without requiring prior knowledge of the identity or the spectra of the analytes.

Another variation using multiple detection is the molecular weight chromatograph,[17] which is a GC with two gas density detectors and two

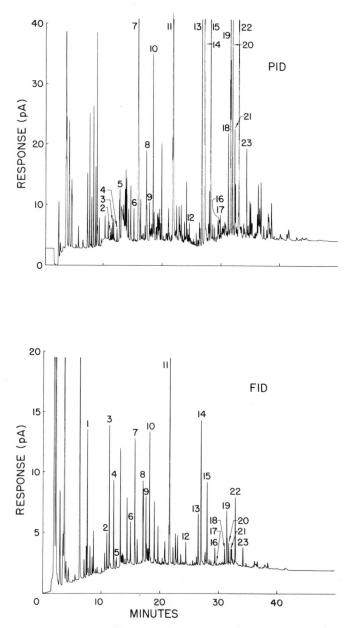

Figure 6.7. Dual channel presentation of GC analysis of air contaminants in parking lot. For identification of peaks, see Table 1. Reprinted with permission from W. Nutmagul, D. R. Cronn, and H. H. Hill, Jr., *Anal. Chem.* **1983,** *55,* 2160. Copyright 1983, American Chemical Society.

TABLE 1   Identification of Peaks in Figure 6.7[a]

Peak No.	Compound	Normalized PID/FID
1	n-Butane	—
2	2,3-Dimethylbutane	4
3	2-Methylpentane	3
4	3-Methylpentane	3
5	1-Hexene	75
6	2,4-Dimethylpentane	15
7	Benzene	129
8	2,3-Dimethylpentane	40
9	3-Methylhexane	12
10	2,2,4-Trimethylpentane	28
11	Toluene	100
12	n-Octane	22
13	Ethylbenzene	84
14	p- and m-Xylene	116
15	o-Xylene	82
16	n-Nonane	25
17	Isopropylbenzene	134
18	n-Propylbenzene	103
19	p-Ethyltoluene	76
20	1,3,5-Trimethylbenzene	158
21	o-Ethyltoluene	105
22	1,2,4-Trimethylbenzene	123
23	1,2,3-Trimethylbenzene	95

[a] Including observed PID/FID ratios normalized to toluene. Reprinted from ref. 12, with permission.

different carrier gases. The data from this instrument can be used to calculate the molecular weight of each analyte in the range from 2 to over 400.

Auxiliary Instruments.   Auxiliary instruments can be used on the fly as special detectors, or analytes can be trapped and taken to other instruments. Instruments that have been used with chromatography include the mass spectrometer (MS), the infrared spectrometer (IR), the nuclear magnetic resonance spectrometer (NMR), the polarograph, the fluorescence spectrophotometer, and the Raman spectrometer, among others. The two most popular ones are MS and IR, and they will be discussed in more detail in Chapter 11. In the beginning of this chapter we noted the utility of GC/MS and LC/MS.

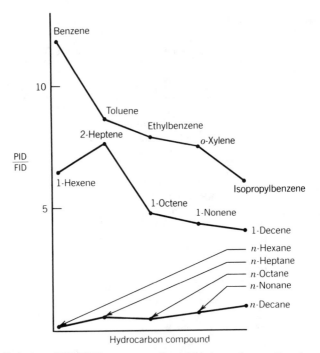

Figure 6.8. Relative (PID/FID) response for 15 hydrocarbons. Reprinted with permission from W. Nutmagul, D. R. Cronn, and H. H. Hill, Jr., *Anal. Chem.* **1983,** *55,* 2160. Copyright 1983, American Chemical Society.

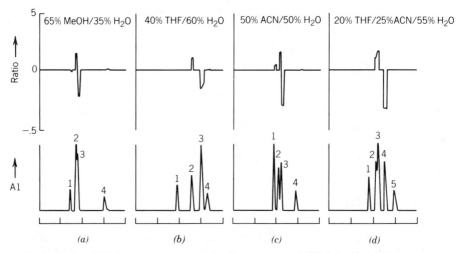

Figure 6.9. LC chromatograms and ratiograms using UV detection at two wavelengths (255 and 280 nm) with four different mobile phases. Reprinted with permission from A. C. J. H. Drouen, H. A. H. Billiet, and L. De Galan, *Anal. Chem.* **1984,** *56,* 971. Copyright 1984, American Chemical Society.

REFERENCES

1. E. S. Kovats, *Helv. Chim. Acta* **1958,** *41,* 1915.

2. G. D. Mitra and N. C. Saha, *J. Chromatogr. Sci.* **1970,** *8,* 95.

3. R. A. Hively and R. E. Hinton, *J. Gas Chromatogr.* **1968,** *6,* 203; L. S. Ettre and K. Billeb, *J. Chromatogr.* **1967,** *30,* 1; N. C. Saha and G. D. Mitra, *J. Chromatogr. Sci.* **1970,** *8,* 84; P. G. Robinson and A. L. Odell, *J. Chromatogr.* **1971,** *57,* 11.

4. W. O. McReynolds, *Gas Chromatographic Retention Data,* Preston Technical Abstracts, Evanston, Ill., 1966.

5. H. Colin and G. Guiochon, *J. Chromatogr. Sci.* **1980,** *18,* 54.

6. V. G. Berezkin, *Chemical Methods in Gas Chromatography,* Elsevier, Amsterdam, 1983; D. A. Leathard in *Advances in Chromatography,* Vol. 13, Giddings (ed.), Dekker, New York, 1975, p. 265; L. S. Ettre and W. H. McFadden (eds.), *Ancillary Techniques of Gas Chromatography,* Wiley-Interscience, New York, 1969.

7. M. Beroza, *Anal. Chem.* **1962,** *34,* 1801.

8. V. G. Berezkin, V. R. Alishoyev, and I. B. Nemirovskaya, *Gas Chromatography of Polymers,* Elsevier, Amsterdam, (reprinted) 1983; R. W. May, E. F. Pearson, and D. Scothern, *Pyrolysis GC,* Chemical Society, London, 1977; J. O. Walker and C. J. Wolf, *J. Chromatogr. Sci.* **1970,** *8,* 513. J. C. Hu, *Adv. Chromatogr. N. Y.,* **1984,** *23,* 149.

9. J. T. Walsh and C. Merritt, Jr., *Anal. Chem.* **1960,** *32,* 1378.

10. R. W. Frei, H. Jansen, and U. A. Th. Brinkman, *Anal. Chem.* **1985,** *57,* 1529A.

11. B. A. Bierl, M. Beroza, and C. W. Collier, *Science* **1970,** *170,* 87.

12. W. Nutmagul, D. R. Cronn, and H. H. Hill, Jr., *Anal. Chem.* **1983,** *55,* 2160.

13. A. C. J. H. Drouen, H. A. H. Billiet, and L. De Galan, *Anal. Chem.* **1984,** *56,* 971.

14. A. C. J. H. Drouen, H. A. H. Billiet, and L. De Galan, *Anal. Chem.* **1985,** *57,* 962.

15. K. Jinno, T. Hoshino, T. Hondo, M. Saito, and M. Senda, *Anal. Chem.* **1986,** *58,* 2696.

16. R. F. Lacey, *Anal. Chem.* **1986,** *58,* 1404.

17. C. E. Bennett, L. W. DiCave, Jr., D. G. Paul, J. A. Wegener, and L. J. Levase, *Am. Lab.* **1971,** 3(5), 67.

The subject of quantitative analysis is intimately tied to detectors, so this chapter begins with a general discussion about detectors. For the most part, the definitions and recommendations about detectors are those specified by the American Society for Testing and Materials (ASTM). In many instances, the general definitions do not apply equally to all detectors, but individual detectors will be discussed in the respective GC and LC chapters, and specific, individual recommendations are included there.

The methods of quantitative analysis presented in this chapter are not unique to chromatography, but their principles will be discussed as they relate to chromatographic measurements. They will be preceded by a short section dealing with conventional statistical definitions and terms.

DETECTORS

The usual definition of chromatography as a method of separation does not imply that a detector is involved. In TLC and PC, detectors are not required for qualitative analysis, and consequently, these simple techniques are used primarily for qualitative screening. The opposite is true for the column methods, and one normally assumes that a *chromatograph* includes a detector. This section deals with the general aspects of detectors like those used in GC and LC column chromatographs.

Classification of Detectors

While there are many ways detectors can be classified, there are five classifications that are especially important in chromatography. Each will be discussed briefly; the first one is probably the most important.

Concentration Versus Mass Flow Rate. The most common type of chromatographic detector is one that produces a signal that is proportional to the concentration (mass/volume) of analyte. Typical examples are given in Table 1. The concentration to which the detector responds is the concentration of an analyte in the detector cell, and therefore the volume of the detector cell is important. Generally small cell volumes are desirable,

TABLE 1   Classification of Some Common Detectors

Concentration Type	Mass Flow Rate Type
Gas Chromatography	
Thermal conductivity (TCD)	Flame ionization (FID)
Electron Capture (ECD)	Other ionization types
	Flame photometric (FPD)
Liquid Chromatography	
UV/Vis Absorption	Mass spectrometry (MS)
Fluorescence	Transport (FID)
Amperometric	
Conductometric	
Refractive index (RI)	

but the size actually depends on the width of the narrowest peak to be measured.

Take, for example, an LC peak whose 4σ width was measured to be 0.80 mm with a chart speed of 1.0 cm/min and a flow rate of 1.0 mL/min. By multiplying these values together, we find that this width corresponds to a volume of 80 μL. If the detector had a volume of 80 μL or more, the entire analyte could be contained in it at one time and the peak would appear to be very broad due to dilution with mobile phase. A smaller detector volume—say 8 μL—would have to be swept 10 times to accommodate all the analyte, and it would give a tall, sharp peak that more nearly represents the actual analyte distribution. As a general rule, the detector volume should be $\frac{1}{4}$ or less of the volume of the smallest peak (usually the first peak unless there is a large solvent peak).

If a data system were being used, it too would need to have a short sampling time in order to produce a sufficiently large number of digital signals to take advantage of the small detector volume. The smaller volume detector could produce at least 10 different signals as the analyte passed, and that would be a sufficient number to get a normal peak shape in most cases.

The other type of detector produces a signal that is proportional to the rate of mass flow (mass/time). It is independent of any detector volume, although connecting tubes are kept small to prevent extracolumn peak broadening. Its signal depends only on the actual mass of analyte passing through it; if the flow stops, the signal stops. The most common ones are also listed in Table 1.

These two types of detector[1] have very different responses to flow changes, as just suggested. Consider an extreme case where the flow in a detector actually stops while the analyte is in the detector, and then it starts again. The peaks that would result are shown in Figure 7.1. When the flow stops, the concentration type has some concentration of analyte present in its volume, and it maintains that signal until the flow is resumed, while the mass flow rate type is not being presented with any new analyte and its signal decays to zero, where it remains until the flow is resumed.

Now consider a less dramatic change in flow—say from 1 mL/min to 0.5 mL/min. Suppose the peak detected at the higher flow had a relative height of 1.0 and a relative area of 1.0 for both types of detector. At the lower flow, the area of the concentration detector would be doubled but its height would remain at 1.0. For the mass flow rate type, the lower flow would result in an unchanged area, and since the peak would be broader, the height would be reduced to 0.5. These relationships are shown in Figure 7.2, and we conclude that the concentration detector produces areas that are flow dependent, but the mass flow rate type does not. On the other hand, the mass flow rate type produces a signal (peak height) that is flow dependent, but the concentration type does not.

These differences are important in quantitative analysis. First of all, it is not fair to compare detectors that are of different types unless the flow rate and concentration are specified. Second, it will be difficult to compare

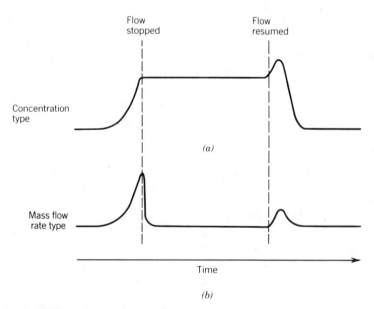

Figure 7.1. Effect of stopping the flow on two types of detector: concentration and mass flow rate.

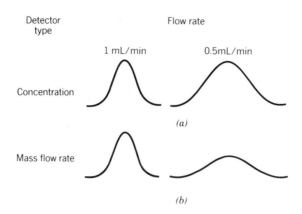

Figure 7.2.   Effect of flow rate on peak sizes for two types of detector: concentration and mass flow rate.

their sensitivities since their signals have different units. Third, since flow might change slightly during a run, concentration detectors could produce erroneous areas and mass flow detectors could produce erroneous peak heights. These topics will be elaborated upon later in this chapter.

Bulk Property versus Specific Property. Detectors referred to as *bulk property* types are constantly measuring a particular property that is exhibited by both the mobile phase and the analyte. For example, a refractive index (RI) detector used in LC is constantly measuring the RI of the mobile phase. When an analyte appears in the detector, there is a change in RI, and it is detected by the transducer and recorded. Often this type of detector has two cells, a sample and a reference cell; the output signal is the difference or ratio between them. These detectors are inherently insensitive because they only measure *changes* in a given property. Many of them cannot be used in gradient elution LC because the bulk property of the mobile phase changes as its composition changes.

By comparison, a specific property type produces no signal (or perhaps only a very small signal) when there is no sample present. The appearance of a sample in the detector introduces a new type of signal and thus produces a relatively large signal (compared to zero signal for the baseline). This type is also called an *analyte property* or *solute property detector*. Some examples are given in Table 2. They are inherently the more sensitive type.

Selective versus Universal. This detector category refers to the number of analytes that can be detected. A universal detector theoretically detects all samples, while the selective type responds to particular types of com-

TABLE 2   Classification of Some Common Detectors

Bulk Property	Specific Property
Gas Chromatography	
Thermal conductivity (TCD)	Flame ionization (FID)
Electron capture (ECD)	Other ionization types
	Flame photometric (FPD)
Liquid Chromatography	
Refractive index (RI)	UV/Vis Absorption
Conductometric	Fluorometric
	Transport (FID)
	Amperometric

pounds or functional groups. Both types have advantages and disadvantages. For qualitative screening, a universal detector will provide information about the total number of analytes, while a selective one will pick out a select group that may aid in its identification. Often the selective type is more sensitive, which is also advantageous for trace analysis. A very crowded, complex chromatogram may be simplified by a selective detector. Both types can be used to advantage in quantitative analysis, but preferably after the sample has been characterized with a universal detector.

Destructive versus Nondestructive. Nondestructive-type detectors are necessary if the separated analytes are to be reclaimed for further analysis, as, for example, when identifications are to be performed using auxiliary instruments. One way to utilize destructive detectors in this situation is to split the effluent stream and send only part of it to the detector, collecting the rest for analysis.

Analog versus Digital. Most detectors produce analog (continuous) signals that must be digitized before they can be manipulated by a digital computer. The main exception is the radioactive detector.

Detector Characteristics

The most important detector characteristic is the signal it produces, of course, and that topic is treated throughout this chapter. In this section we will define two other important characteristics, noise and time constant.

Noise. Noise is the signal produced by a detector in the absence of a sample. Usually it is given in the same units as the normal detector signal—volts, amps, absorbance units, and so on. It is caused by the electronic components from which it is made, from stray signals in the environment, and from contamination. Circuit design can minimize noise from the former source, shielding and grounding can help remove environmental noise, and sample pretreatment can often remove contaminants before the sample is chromatographed.

Figure 7.3 is taken from an ASTM recommendation for a GC detector[2] and shows a typical noisy baseline. As indicated, the noise is defined as the detector signal range (in this case in mV) between the two parallel lines that enclose the random fluctuations. In some other specifications, noise is further categorized as short term (0.5 to 1 min) and long term (10 min). In addition, the figure shows a long term *drift* over a period of 30 min. Sometimes drift is referred to as long term noise.

The level of noise restricts the minimum signal that can be detected and attributed to an analyte, so it is important to keep it to a minimum. A detector characteristic that is often more meaningful than the noise is the ratio of the signal-to-noise, *S/N*. In most chromatographic work it is

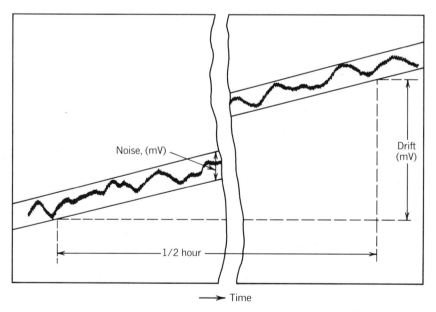

Figure 7.3. Example for the measurement of noise level and drift for a TCD. Copyright ASTM. Reprinted with permission.

agreed that the smallest signal that can be attributed to an analyte is one whose S/N is 2.

Time Constant. The time constant τ is a measure of the speed of response of a detector. Specifically, it is the time (usually in seconds or milliseconds) a detector takes to respond to 63.2% of a sudden change of signal, as shown in Figure 7.4. The full response (actually 98%) takes four time constants and is referred to as the *response time*. Unfortunately, some workers define *response time* as 2.2 time constants (not 4.0), corresponding to 90% of full scale deflection (not 98%); others define a *rise time* as the time for the signal to rise from 10 to 90%. To further confuse the situation, some use the terms *time constant* and *response time* interchangeably. This lack of consistency can be found in the ASTM specifications.

Nevertheless, Figure 7.5 shows the effect of increasingly longer time constants in distorting the shape of a chromatographic peak. The deleterious effects on chromatographic peaks are the changes in retention time and peak width, both of which get larger as the time constant gets larger. The area is unaffected, so quantitative measurements based on area will still be accurate, but those based on peak height will be in error.

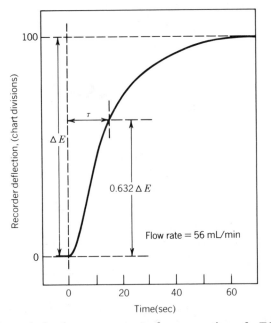

Figure 7.4. Example for the measurement of response time of a TCD. Copyright ASTM. Reprinted with permission.

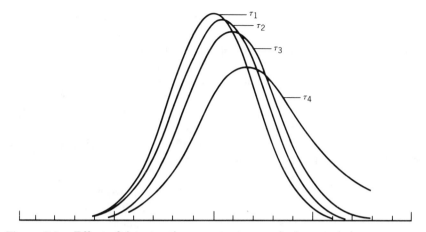

Figure 7.5. Effect of detector time constant on peak characteristics; $\tau_1 < \tau_2 < \tau_3 < \tau_4$.

One recommendation for LC detectors[3] is that the time constant should be less than 10% of the peak width at half height, w_h. Thus, a peak width of 50 μL at a flow rate of 1 mL/min corresponds to a time constant of 0.3 sec. This is the order of magnitude needed for most chromatographic detectors.

One advantage of a large time constant is *decrease in short term noise*, which is also called *damping*. The temptation to improve one's chromatogram by increasing the time constant to decrease the noise must be avoided. Also, consideration must be given to the time constants of all components in the detector network; for example, the recorder must have a speed comparable to the detector itself.

Detector Specifications

The two detector characteristics just discussed, noise and time constant, are important detector specifications too. However, this section will focus on the detector signal and its relationship to quantitative analysis. There are slight but significant differences between the specifications for concentration and mass flow rate types of detectors that will necessitate some duplication in the discussion. Recall that the concentration detector signal is proportional to concentration (e.g., g/mL) and the mass flow rate detector to mass flow (e.g., g/sec).

Sensitivity. *Sensitivity* is a measure of the amount of signal generated by the detector for a given amount of analyte, and it should be a constant. The signal can have a variety of units as already indicated—volts, amperes, absorbance units, and so on. If the signal is in millivolts, the concentration detector will have its sensitivity given in mV/(conc) or mV/(g/mL), which is mV mL/g. The mass flow rate detector will have its sensitivity given in mV sec/g. If the signal is in amperes, the respective units are A mL/g and C/g (a coulomb is an ampere-second).

Detector signals are usually determined for more than one mass of analyte, resulting in a graph like that shown in Figure 7.6. Note that the line deviates from linearity at the high end and that the slope of this graph has the units of sensitivity for a concentration detector, mV mL/mg. Because the range of analyte sizes often extends over several orders of magnitude, this plot is often done on a log–log basis. Unfortunately, this has the effect of covering up some of the deviations from linearity and is overly optimistic. A better graph is the one shown in Figure 7.7, which is taken from an ASTM specification.[2] Here the analyte mass can be on a log basis, but the ordinate is sensitivity, not signal, and it is on a linear scale.

Detectivity. The smallest concentration plotted in Figure 7.7 is the *minimum detectable quantity*, MDQ, or the *detectivity*. This limit was defined earlier as twice the noise; MDQ is a signal S equal to 2N. Since the detector noise is usually specified in the same units as the signal, the detectivity can carry these same units. For example, if the signal is in

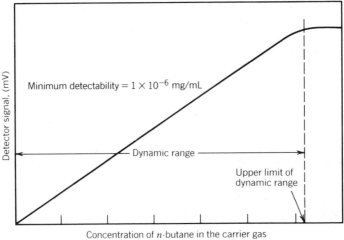

Figure 7.6. Example of a plot to determine the dynamic range of a TCD. Copyright ASTM. Reprinted with permission.

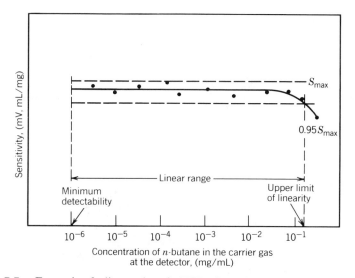

Figure 7.7. Example of a linear plot of a TCD. Copyright ASTM. Reprinted with permission.

millivolts, the detectivity can be given in millivolts. More often, however, the detectivity of a concentration detector will be in concentration units, (g/mL), and that of a mass flow rate detector in grams per second. These units are obtained by dividing the minimum detectable signal by the sensitivity; for example,

$$\text{MDQ} = \frac{\text{mV}}{\text{mV mL/g}} = \frac{\text{g}}{\text{mL}} \qquad (1)$$

However, a detectivity of 10^{-9} g/mL for a concentration detector means that this concentration is the smallest one that can be detected *at that concentration in the detector cell*. Consequently if a sample of this concentration is introduced into a chromatograph and run, its concentration will be somewhat less when it reaches the detector, because we have seen that an analyte is diluted and its zone becomes wider as it passes through the column. Thus, the amount of analyte that can actually be run and detected can differ from the MDQ. This situation has produced other terms like *minimum detectable concentration*, MDC, which will depend on the peak width (in milliliters). Analogously, for the mass flow rate detectors, the *minimum detectable mass*, MDM, will depend on the peak width in time units. In both cases, the quantity to be injected depends on the final peak width and can be calculated as

$$\text{amount to inject} = \text{MDQ} \times w_h \qquad (2)$$

where w_h is the width at half-height and has the units of volume for the concentration detector and time for the mass flow rate detector. In both cases, the units on the "amount to inject" will be mass (g).

Linearity. Figure 7.7 shows that the sensitivity falls off at some value on the high side. To establish the linearity of a detector this upper limit must be specified. The ASTM recommendation, as shown in the figure, is that the upper limit is the analyte size (concentration in the figure) corresponding to a sensitivity equal to 95% of the maximum sensitivity. The upper dashed line in the figure is drawn through the point representing the maximum sensitivity, and the lower dashed line is 0.95 of that value.

Having established the ends of the linear range, the MDQ and the upper limit, the linearity is defined as

$$\text{linearity} = \frac{\text{upper limit}}{\text{MDQ}} \qquad (3)$$

which is dimensionless. Obviously, a large value is desired for this parameter.

In all of these plots and discussions, it has been the detector *signal* that has been used. However, for quantitative analysis, the *area* is more commonly measured than the signal (peak height). So, for quantitative analysis, a calibration graph will look like Figure 7.6 but the variables on the axes will be different. The ordinate can be height or area, and the abscissa can be mass or concentration. The same linear relationships should hold no matter how the graph is plotted. Usually we assume that a peak can be approximated by a triangle and its area as some product of the peak height and width, and since width does not change for a given analyte under given conditions, area and peak height should be directly proportional. Of course, changes in flow rate can cause errors in peak width for a concentration detector or in peak height for a mass flow rate detector, as we have seen.

STATISTICS OF QUANTITATIVE CALCULATIONS

A brief discussion of some statistical measurements used in quantitative analysis is necessary before proceeding. When a measurement is made, there is some random error associated with it. This error is called *inde-*

terminate error to distinguish it from *determinate error*—error that is not random and should be found and corrected. We will deal only with the former, which, since it is random, can be compensated for by making a large number of measurements. Some errors will be positive and some negative, so large numbers of measurements will tend to cancel each other. If a large enough number of measurements are made, the distribution of values will approximate a Gaussian curve, which can be characterized by two variables—the *central tendency* and the *variation about the central tendency*. Two measures of the central tendency are the mean (or average), $\overline{X}$, and the median. One of these values is usually taken as the "correct" value for an analysis, although, statistically, it is the most probable value since there really is no "correct" value. The ability of an analyst to perform an analysis and determine this "correct" value is referred to as the *accuracy*.

The spread of data about the mean is usually measured with the standard deviation σ, but another common parameter is the range. (In our discussion of Gaussian chromatographic peaks, we have used σ as $\frac{1}{4}$ of the peak width.) By definition,

$$\sigma = \sqrt{\frac{\sum(X - \overline{X})^2}{(n - 1)}} \tag{4}$$

and the square of σ is called the *variance*. A good analyst working with a good method will produce data that have a small standard deviation, and this is referred to as the *precision*.

Precision and accuracy can be easily represented as shots at a target, as shown in Figure 7.8. Figure 7.8*a* shows good accuracy and precision;

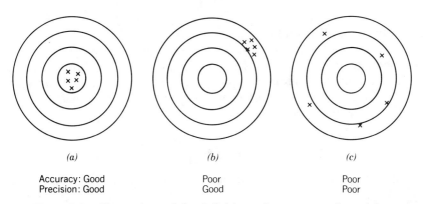

(a)	(b)	(c)
Accuracy: Good	Poor	Poor
Precision: Good	Good	Poor

Figure 7.8.   Illustrations of the definitions of accuracy and precision.

Figure 7.8*b* good precision but poor accuracy; and Figure 7.8*c* poor precision that will result in poor accuracy unless a large number of shots are taken. The situation in Figure 7.8*b* suggests that a determinate error is present; maybe the gunsight is out of alignment.

Two other terms are in common use to distinguish two types of precision. One is *repeatability*, which refers to the precision in one lab, by one analyst, and on one instrument. The other is *reproducibility*, which refers to the precision among many labs and consequently many analysts and many instruments. We expect and we find that the data from many labs are more variable than they are from one lab; that is, reproducibility is not as good as repeatability.

In a set of data, a *relative standard deviation* carries more information than the standard deviation itself. The relative standard deviation, or coefficient of variation, is

$$\sigma_{rel} = \frac{\sigma}{\overline{X}} \tag{5}$$

The minimum information usually given to convey the results of an analysis is one of each of the two variables we have discussed—usually the mean and the relative standard deviation.

Before leaving statistics, it will be helpful to note that the correct statistical way to add together several standard deviations is *to add their variances and take the square root*. Thus the total standard deviation σ_T of a multistep process is

$$\sigma_T = \sqrt{\sum_i \sigma^2} \tag{6}$$

Use is made of this relationship in chromatography in calculating the total zone spreading that results from multiple sources. To get the total zone width σ, the variances of the individual sources of zone spreading are summed and the square root is taken. For example, if a chromatograph has appreciable volumes in the tubing used to connect the column to the detector and to the injection valve, these will add to the total width of the peak. The total peak width σ_T is

$$\sigma_T = \sqrt{\sigma_{col}^2 + \sigma_{tub}^2} \tag{7}$$

In later chapters more will be presented on this extracolumn zone spreading.

QUANTITATIVE METHODS

Interlaboratory studies that have been made on a common sample have shown that the errors associated with quantitative analysis by chromatography are much larger than expected. One fairly recent study is typical.[4] Two samples were prepared and sent to 78 labs for analysis by reverse phase LC. When the data were analyzed, it was found that:

The data from five laboratories had to be rejected, statistically.

Of the 700 calculations that were reported, 5% were found to be in error.

A large plate number does not necessarily produce the best quantitative results.

The relative standard deviation of the results from the acceptable laboratories was 3.1 to 4.6%.

These are very sobering findings and should caution chromatographers to exercise care in performing quantitative analyses!

Considerations in Setting Up a Method

So many factors must be considered in setting up a method that the topic cannot be treated adequately here; only a few important issues will be raised.

First of all, a qualitative analysis must precede a quantitative analysis. Perhaps it need not be a complete qualitative analysis, but it is helpful to know as much as possible about the sample. Standards must be available for the analytes to be determined. The sample must be representative and it must be stable. A good chromatographic separation is highly desirable, and the detection system must have the ideal specifications we have discussed. Decisions have to be made about the number of analytes to be determined in the sample, the level of precision that will be needed, and so on. And finally, the choice needs to be made whether to measure peak height or peak area.

Peak Height versus Peak Area. This topic has received a lot of attention.[5] Several fundamental considerations are:

1. Area measurements are generally considered to represent ideally the quantity of analyte being measured. The area represents the integration of the entire sample in the peak, not just the average value.

2. The process of making area measurements is more likely to be in error unless electronic methods and computers are used. Undoubtedly many early GC analyses suffered from poor integration methods, but area measurements should now be as accurate as height measurements.

3. If the mobile phase flow rate is not constant, the concentration-type detectors should show greater errors in area measurements, and mass-flow-rate-type detectors should show greater errors in peak height measurements, for the reasons given earlier in this chapter. Apparently this is a common source of error.

Recent comparisons have been made for LC analyses, usually with UV detectors that are of the concentration type.[4,5] One would expect that area measurements would be less accurate if flow rates varied. Indeed, the 1981[4] study found that peak height measurements were better, although the committee could not resist stating that "in the hands of competent chromatographers the choice of peak height vs. peak area measurements may be a stand-off."[4] A later study[6] among fifty laboratories found that the choice depends on the quality of the chromatographic separation: for good separations, peak areas were as good as or better than peak heights, but for poor separations (overlapped peaks) peak heights were better. A newer report from the same group,[7] using mixtures that were easier to separate and less widely varying in concentration, concluded that peak area was indeed better than peak height.

In summary, peak area is the preferred measurement especially if there are any changes in chromatographic conditions, such as partition ratio, temperature, or sample introduction method, that can cause changes in peak height or width (but not area). However, peak height measurements are less affected by overlapping peaks, noise, and sloping baselines. It must also be remembered that the commonly used concentration detectors are flow sensitive and prone to errors if areas are used for quantitative analysis.

Measurement of Areas. The process of measuring peak areas consists of integrating the area under the peaks. This process also converts an analog signal into a digital signal, which is necessary for computer calculations. Peak area measurements will probably be in error by at least 2% if made manually or mechanically, which puts a rather high limit on the precision attainable. Therefore, electronic or computer methods are a necessity for high precision work; in addition, they make it possible to handle quantitative measurements where the chromatography is poor.

A few examples of the capabilities of automatic data stations are shown in Figure 7.9. The slope sensitivity can be set to distinguish between the

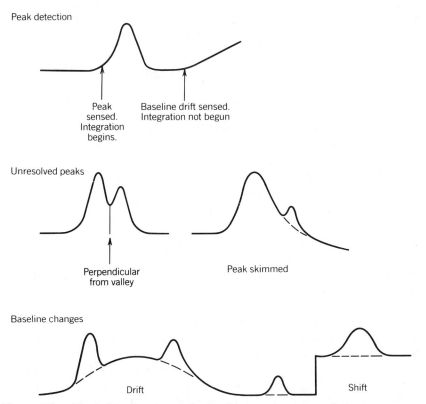

Figure 7.9. Examples of some methods of integrating areas of chromatograms with a computer data station.

start of a peak and the start of a sloping baseline (which might result in programmed temperature GC). The area under unresolved peaks can be separated by dropping a perpendicular from the valley between them. Alternatively, a small peak on the tailing edge of a larger peak can be *skimmed off*. Baseline changes can be compensated for by forcing new baselines to be used. Proper measurement of peak areas requires considerable attention to these measurement options. When properly done, the error associated with the measurement should be well under 1%.

Methods of Quantitative Analysis

Five methods of quantitative analysis will be discussed briefly, proceeding from the most simple and least accurate to those which are capable of higher accuracy.

Area Normalization. As the name implies, area normalization is really a calculation of area percentage. If X is the unknown analyte,

$$\%X = \left(\frac{A_x}{\sum_i A_i}\right) \times 100 \tag{8}$$

where A_x is the area of X and the denominator is the sum of all the areas. For this method to be accurate, the following criteria must be met:

1. All analytes must be eluted.
2. All analytes must be detected.
3. All analytes must have the same sensitivity.

It is not very likely that all three conditions will be met, but this method is very simple and is often useful if a semiquantitative analysis is sufficient or if some analytes have not been identified or are not available in pure form for standards.

Area Normalization with Response Factors. If standards are available, the third limitation can be removed by running the standards to obtain correction factors or so-called relative response factors f. One substance (it can be an analyte in the sample) is chosen as the standard, and its response factor f_s is given an arbitrary value like 100. Mixtures, by weight, are made of the standard and the other analytes, and they are chromatographed. The areas of the two peaks—A_s and A_x for the standard and the unknown, respectively—are measured, and the relative response factor of the unknown, f_x, is calculated:

$$f_x = f_s \times \left(\frac{A_s}{A_x}\right) \times \left(\frac{w_x}{w_s}\right) \tag{9}$$

w_x/w_s is the weight ratio of the unknown to the standard. Relative response factors of some common compounds have been published for the two most common GC detectors,[8] and a variety of other relative response factors can be found in the literature. For the highest accuracy, one should determine his/her own factors, however.

When the unknown sample is run, each area is measured and multiplied by its factor. Then, the percentage is calculated as before:

$$\%X = \left[\frac{(A_x f_x)}{\sum_i (A_i f_i)}\right] \times 100 \tag{10}$$

External Standard. This method is usually performed graphically. Known amounts of the analyte of interest are chromatographed, the areas are measured, and a calibration curve like Figure 7.6 is plotted. If the standard solutions of analyte vary in concentration, a constant volume must be introduced to the column for each. This requires a reproducible method of sample introduction; a valve is adequate, but syringe injection in GC is usually inadequate, particularly for syringes that contain sample in the needle. Errors around 10% are common.

If a calibration curve is not made and a data system is used to make the calculations, a slightly different procedure is followed. A calibration mixture prepared from pure standards is made by weight and chromatographed. Absolute calibration factors, equal to the grams per area produced, are stored in the data system for each analyte. When the unknown mixture is run, these factors are multiplied times the respective areas of each analyte in the unknown resulting in a value for the mass of each analyte. This procedure is a one-point calibration, as compared to the multipoint curve described before, and is somewhat less precise. Note also that these calibration factors are not the same as the relative response factors used in the area normalization method.

Internal Standard. This method and the next one are particularly useful with techniques like chromatography, which are not too reproducible from day to day, and in situations where one does not want to recalibrate often. It does not require exact or consistent sample volumes or response factors since the latter are built into the method. The standard chosen for this method cannot ever be a component in a sample; a known amount of this standard is added to each sample; hence the name *internal* standard. It must meet several criteria:

1. It should elute near the peaks of interest, but
2. It must be well resolved from them.
3. It should be chemically similar to the analytes of interest and not react with any sample components.
4. Like any standard, it must be available in pure form.

The standard is added to the sample in about the same concentration as the analyte(s) of interest and prior to any chemical derivatization or other reactions. If many analytes are to be determined in one sample, several internal standards may be used in order that the preceding criteria are met.

One or more calibration mixtures are made from pure samples of the analytes, depending if a one-point calibration is desired or if a graph is to be plotted (see discussion under the external standard method). A

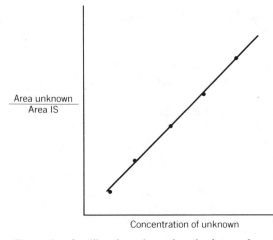

Figure 7.10.   Example of calibration plot using the internal standard method.

known amount of internal standard is added to the calibration mixture(s) and to the unknown. Usually the same amount of standard is added to each, and this is often done most conveniently by volume. All areas are measured and referenced to the area of the internal standard, either by the data system or by hand. If multiple standards are used, a calibration graph like that shown in Figure 7.10 is plotted where both axes are relative to the standard. If, as is shown in the figure, the same amount of internal standard is added to each calibration mixture and each unknown, the abscissa can simply represent concentration, not relative concentration. The unknown is determined from the calibration curve or from the calibration data in the data station. In either case, any variations in conditions from one run to the next are cancelled out by referencing all data to the internal standard. In principle, this method should produce better accuracy, but some exceptions have been reported.[9]

Standard Addition Method.   In this method the standard is also added to the sample, but the chemical chosen as the standard is the *same* as the analyte of interest. It requires a highly reproducible sample volume, which is a limitation with syringe injection in GC, as noted earlier.

The principle of this method is that the extra signal produced by the addition of standard is proportional to the original signal. Equations can be used to make the necessary calculations, but the principle is more easily seen graphically. Figure 7.11 shows a typical standard addition calibration plot. Note that a signal is present when no standard is added; it represents the original concentration, which is to be determined. As

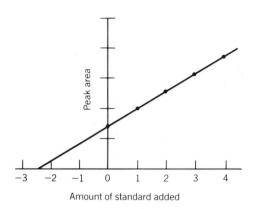

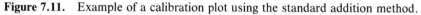

Amount of standard added

Figure 7.11.   Example of a calibration plot using the standard addition method.

increasing amounts of standard are added to the sample, the signal increases, producing a straight line calibration. To find the original "unknown" amount, the straight line is extrapolated until it crosses the abscissa; the absolute value on the abscissa is the original concentration. In actual practice, the situation is more complex, and a thorough summary has been provided by Bader.[10] For example, different calculations are required when the total volume is kept constant and when it varies as standard is added.

Matisova and co-workers[11] have suggested that the need for a reproducible sample volume can be eliminated by combining the standard addition method with an *in situ* internal standard method. In the quantitative analysis of hydrocarbons in petroleum, they chose ethyl benzene as the standard for addition, but they used an unknown neighboring peak as an internal standard to which they referenced their data. This procedure eliminated the dependency on sample size and provided better quantitation than the area normalization method they were using.

REFERENCES

1. I. Halasz, *Anal. Chem.* **1964,** *36,* 1428.
2. ASTM E 516-74, *Standard Recommended Practice for Testing Thermal Conductivity Detectors Used in Gas Chromatography,* American Society for Testing and Materials, Philadelphia, 1974.
3. E. L. Johnson and R. Stevenson, *Basic Liquid Chromatography,* Varian Associates, Palo Alto, Calif., 1978, p. 278.
4. Subcommittee E-19.08 Task Group, *J. Chromatogr. Sci.* **1981,** *19,* 338.

5. W. Kipiniak, *J. Chromatogr. Sci.* **1981**, *19*, 332.

6. R. W. McCoy, R. L. Aiken, R. E. Pauls, E. R. Ziegel, T. Wolf, G. T. Fritz, and D. M. Marmion, *J. Chromatogr. Sci.* **1984**, *22*, 425.

7. R. E. Pauls, R. W. McCoy, E. R. Ziegel, T. Wolf, G. T. Fritz, and D. M. Marmion, *J. Chromatogr. Sci.* **1986**, *24*, 273.

8. W. A. Dietz, *J. Gas Chromatogr.* **1967**, *5*, 68.

9. P. Haefelfinger, *J. Chromatogr.* **1981**, *218*, 73.

10. M. Bader, *J. Chem. Educ.* **1980**, *57*, 703.

11. E. Matisova, J. Krupcik, P. Cellar, and J. Garaj, *J. Chromatogr.* **1984**, *303*, 151.

GAS CHROMATOGRAPHY **8**

Before GC became popular in the late 1950s, the only way to separate volatile materials was by distillation, which separates materials based on differences in vapor pressure or boiling point. GC is similar in that respect, but its separations also depend on the nature of the stationary phase, which gives it much more versatility than distillation. Imagine the pleasure and surprise of distillers who could now separate materials with close boiling points, like benzene and cyclohexane. And it was easy, fast, and not too expensive; in addition, they didn't have to worry about azeotropes.

Let us look at the benzene–cyclohexane separation more closely as we summarize how GC works. The boiling points of benzene and cyclohexane are nearly the same, 80.1 and 81.4°C respectively. Any GC separation will have to depend on differences in the intermolecular interactions between the stationary phase and these two analytes, both of which are nonpolar hydrocarbons. What differences could be exploited with GC? Benzene has a π-electron cloud, which should make it more susceptible to induction effects and perhaps dispersion attractions (Chapter 3). Therefore we should choose a stationary liquid phase that would accentuate this difference—a polar one; also, using the "like-dissolves-like" rule we might choose an aromatic compound that would interact more with benzene than with cyclohexane. One possible liquid phase that meets these criteria is dinonylphthalate, and it has been used to separate benzene and cyclohexane. The relative retention has been found to be 1.6, which represents a very good separation.[1]

One way of expressing the extent of the interaction between these analytes and the liquid phase is by activity coefficients. According to Raoult's law, the partial pressure of a solute like cyclohexane in a solvent like dinonylphthalate is given by

$$p_{cy} = X_{cy}\gamma_{cy}p_{cy}^0 \tag{1}$$

where p_{cy} is the partial pressure of cyclohexane, X_{cy} is its mole fraction, γ_{cy} is its activity coefficient, and p_{cy}^0 is the vapor pressure of pure cyclohexane. The relative retention of the two solutes is also equal to the ratio of their partial pressures, and since their mole fractions and pure vapor pressures would be about equal,

$$\alpha = \frac{(V_R')_{ben}}{(V_R')_{cy}} = \frac{\gamma_{cy}}{\gamma_{ben}} = 1.6 \qquad (2)$$

Activity coefficients can be calculated from GC data according to Eq. (3):

$$\gamma = \frac{1.7 \times 10^5}{V_g p^0 MW} \qquad (3)$$

where MW is the molecular weight of the stationary liquid and V_g is the *specific* retention volume (the net retention volume at 0°C per gram of liquid phase). In this example, the activity coefficients at 326 K were found to be 0.52 for benzene and 0.82 for cyclohexane, respectively,[1] and their ratio is 1.6, in agreement with Eq. (2). Remember, the more the activity coefficient deviates from 1, the greater is the interaction between the solute and the stationary phase.

In other GC separations, the differences in vapor pressure, or a combination of the two effects used to produce the desired separation, can be exploited. We saw in Chapter 6 how the members of a homologous series are separated according to differences in boiling points.

The earliest columns were packed with solids or with solid supports coated with liquids. In accordance with our naming convention, these two types of chromatography are called gas–solid chromatography (GSC) and gas–liquid chromatography (GLC). Of the two, the latter is more commonly used, and most of this chapter concerns GLC.

In 1957 Golay published his ideas for using columns that were not packed but were open tubes.[2] These tubes had to have small inside diameters so they became known as *capillary* columns, but the name *open tubular (OT) columns* is more descriptive and preferred. These two types of column necessitate slightly different chromatographic instruments, but the discussion that follows will attempt to integrate them together for simplicity of presentation. In the United States, packed columns were much more widely used than OT columns until recently; consequently many laboratories have packed column instruments that, since they will not accept OT columns without modification, are still in use even though OT columns would be preferable for many of their separations. Conversion kits and columns with characteristics intermediate between the two extremes are currently popular.

INSTRUMENTATION

The essential parts of a gas chromatograph as shown in Figure 8.1 are: carrier gas, flow or pressure regulator, injection port or valve, column, and detector. Usually there are three heated zones, each separately controlled, for the inlet area, the column, and the detector. Connections between these heated zones must also be kept hot enough to prevent condensation of analytes in them. Chromatographs designed for OT columns usually have additional features: a more elaborate injection port that allows for sample splitting and a provision for some additional "make-up" gas for the detector. Information about commercial instruments can be found in the review by Bayer.[3]

The Mobile Phase

The most popular carrier gases are nitrogen, helium, and hydrogen. They must be very pure, and they are chosen for their inertness since, as we have seen, their only purpose is to carry the analyte vapors through the column. Helium is the most popular in the United States because of its higher efficiency at faster flow rates, but it is expensive and in limited supply. Hydrogen is increasing in popularity, especially for OT columns. Sometimes the choice of carrier is dictated by the detector.

The pressure drop across a packed column can range from 50 to 300 kPa* and is much less for OT columns. It is regulated with one or more

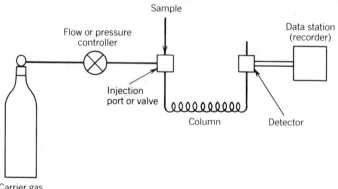

Figure 8.1.   Components of a gas chromatograph.

* The standard unit of pressure in the SI is the pascal, Pa. For those familiar with old units, the following conversions are given: 1 bar = 100 kPa; 1 atm = 101.3 kPa; 1 torr = 133 Pa; 1 psi = 6.9 kPa.

valves in order to get the desired flow rate (linear velocity). We have already seen that a constant flow rate is desirable so that retention times will not vary and flow sensitive detectors will not become nonlinear. In the technique of programmed temperature (PT) GC discussed later, the viscosity of the carrier gas increases markedly during a run, so a good differential flow controller is required. On the other hand, the split injectors used with OT columns require constant *pressure* regulation, with the unfortunate result that the flow cannot be easily controlled during a PT run.

Injection Ports and Valves

Sample introduction is most often accomplished with a microsyringe through a self-sealing rubber septum, as shown in Figure 8.2. However, valves are more reproducible and are preferred for gases (which require volumes on the order of milliliters—volumes more easily handled with valves). Newer valves are available that can handle the smaller microliter sizes needed for liquids and solutions. In all cases, the objective is to get the sample into the column in as small a volume as possible.

Injection ports for packed columns are aligned so that the sample can be deposited on a heated surface just before the column or directly on the end of the column. In the first instance, the injection port is heated above the boiling point of the sample in order to get rapid volatilization, but for on-column use, the injection port is kept at column temperature. On-column injection is usually preferred because there is less chance of decomposition and the sample is not exposed to a high injection port temperature. Remember also that a typical GC analyte will have a retention ratio of 0.25 or much less. This means that 75% or more of it is sorbed in the stationary phase, and that is where the on-column technique puts it—in the stationary phase. It is probably for this reason that on-column injection is so efficient. The use of large volumes of solvents will wash

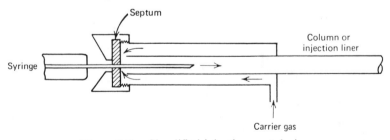

Figure 8.2. Simplified injection port design.

off some of the stationary liquid over a period of time, of course, requiring repacking of the end of the column.

Injection ports for OT columns are usually much more complex because OT columns cannot accept very large samples. By using different liners inside the body of the port and different conditions, a variety of modes of sample introduction can be accommodated with one design. A typical one is shown in Figure 8.3. It is set up for split injection; the sample is deposited in the region designated S, where it is volatilized.

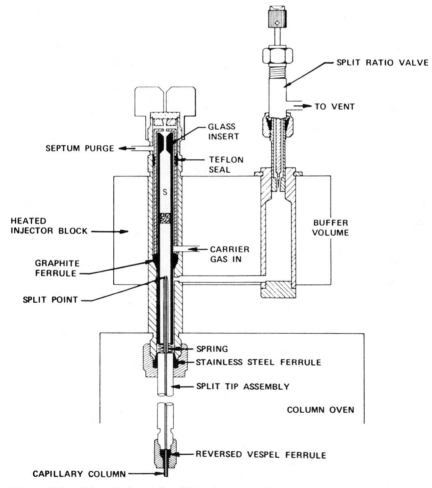

Figure 8.3. Inlet designed for OT columns and sample splitting. Courtesy of Varian Instruments.

Most of it exits through the side tube whose flow is regulated with the needle valve shown. As with most splitters, care must be taken to avoid fractionation of the sample in the splitting process. A plug of glass wool or other inert material in the vaporization region promotes complete volatilization. The buffer volume allows the sample to pass the point of column splitting before reaching the needle valve, which could change the split ratio. Only a small portion of the sample, 0.1 to 10%, is carried into the column, which is shown extending into the splitter. Note also that a small flow of carrier gas purges the septum to keep septum degradation products from contaminating the sample.

The other common mode of sample introduction requires the analyte concentration to be small compared to the large volume of solvent. It is called the *Grob splitless injector*, after its inventor. Usually an open liner is used and the split valve is closed; the column is kept "cold." A relatively large sample of several microliters is injected slowly, and the sample is allowed to evaporate for 30 to 60 sec. Then, the split valve is opened, and the column is programmed to higher temperatures. Using this technique, most of the solvent is vented, giving good sensitivity. A summary comparison of the two methods is given in Table 1.

Many other special devices and procedures have been reported. For example, a special injector is commercially available to permit direct on-column injection for OT columns. One final variation is an injection port that allows the sample to be injected simultaneously on two columns. One recent example of its use is in the analysis of PCBs.[4]

Columns

Columns are usually made of metal (stainless steel, copper, or aluminum) or glass (including the new fused silica OT columns). Inertness is of prime importance, so glass and silica have become increasingly popular, but

TABLE 1   Comparison of Two Common Modes of Sample
Introduction on OT Columns

	Split Mode	Splitless Mode (Grob)
Liner	Contains frit or glass wool	Open
Liner diameter	4 mm	2 mm
Injection method	Simple injection	Several steps (see text)
Sample size	0.1–10 μL	1–10 μL
Sample/solvent ratio	1:1 to 1:1000	$1:10^3$ to $1:10^6$

stainless steel is still used for many packed column applications because it is easier to handle and coil.

The column is either kept at a constant temperature (isothermal GC) or programmed during the run (PTGC). It is a common misconception that the column temperature should exceed the boiling point of the sample in order to keep the analytes in their vapor phases. Actually the column will produce better separations if the temperature is below the sample's boiling point in order to increase the interaction with the stationary phase. The situation desired inside the column can be described in comparison to the water vapor in our environment; there is plenty of vapor well below the boiling point (as we know from those days with high humidity), but we must be above the "dew point" of the analyte or else it will "rain" in the column. The smaller the amount of stationary phase, the lower the temperature at which we can operate, so OT columns are usually run at lower temperatures than packed columns.

Packed Columns. If the stationary phase is a liquid, it is held in the column on an inert *solid support*, which will still appear dry. This support, or an active solid, is the material packed in the column. The nature of these packed beds was discussed in Chapter 2. Theory predicts that improved performance should result from the use of small particles, so some attempts have been made to pack them into columns. Because the diameter of these columns is usually small too, they have been called *packed capillaries* or, more generally, *micropacked columns*. Packed columns with a d_p/d_c ≤0.3 have been put in this classification and reviewed.[5] Some very high efficiencies have been obtained, but sometimes at the expense of very high inlet pressures.

Solid Supports. The particles should be small and uniform; a common mesh range is 100 to 120 (Table 2). They should have a large surface area,

TABLE 2   Mesh and Particle Sizes

Mesh Range	Particle Diameter (μm) From	To	Range (μm)
45/60	354	250	104
60/80	250	177	73
80/100	177	149	28
100/120	149	125	24
325/400	44	37	7

so the most common ones are made from diatomaceous earth, which has a very porous structure. The properties of some common supports are listed in Table 3. The surfaces of the diatomaceous earth types are too active for many applications, and they can be treated to remove most of the activity. The treatments involve acid washing (AW) and silanizing to react with the surface hydroxyl groups.[6] This is necessary even for the inert white supports (e.g., Chromosorb W), and it results in a superior material that is often sold under special trade names, such as Supelco-port™, Chromosorb W-HP™, Gas Chrom Q II™, and Anachrom Q™. The liquids coated on these supports are the same as those used in OT columns, and this topic will be discussed later. However, it is important to note that these silanized supports become very hydrophobic and are not easily wetted by polar liquid phases, so they may not be satisfactory in such cases.

Active Solids (GSC). Most of the common active solids have been used in GSC: silica gel, alumina, charcoal, and so on. Because their surfaces tend to be heterogeneous in activity, they produce undesirable tailing peaks and are not very popular; this problem can be greatly reduced by coating a thin layer of liquid on the support, as is shown in Figure 8.4 for the graphitized carbon support called Carbopack.

TABLE 3   Packings for GC

Name	Manufacturer[a]	Surface Area (m²/g)	Packed Density (g/cm³)	Pore Size (μm)	Maximum % Liquid Phase
Diatomaceous Earth					
Chromosorb P	J	4.0	0.47	0.4–2	30
Chromosorb W	J	1.0	0.24	8–9	15
Chromosorb G	J	0.5	0.58	NA[b]	5
Chromosorb A (prep use)	J	2.7	0.48	NA	25
Fluorocarbon Polymer					
Teflon T-6	D	7.8	0.49	None	10
Glass					
Microbeads	—	0.01	1.4	None	3
Textured beads	C	0.04	1.35	NA	0.5

[a] Code: J = Johns–Manville; D = DuPont; C = Corning Glass (discontinued).
[b] NA = not available.

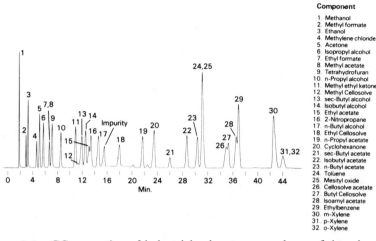

Component

1. Methanol
2. Methyl formate
3. Ethanol
4. Methylene chloride
5. Acetone
6. Isopropyl alcohol
7. Ethyl formate
8. Methyl acetate
9. Tetrahydrofuran
10. n-Propyl alcohol
11. Methyl ethyl ketone
12. Methyl Cellosolve
13. sec-Butyl alcohol
14. Isobutyl alcohol
15. Ethyl acetate
16. 2-Nitropropane
17. n-Butyl alcohol
18. Ethyl Cellosolve
19. n-Propyl acetate
20. Cyclohexanone
21. sec-Butyl acetate
22. Isobutyl acetate
23. n-Butyl acetate
24. Toluene
25. Mesityl oxide
26. Cellosolve acetate
27. Butyl Cellosolve
28. Isoamyl acetate
29. Ethylbenzene
30. m-Xylene
31. p-Xylene
32. o-Xylene

Figure 8.4. GC separation of industrial solvents on a column of the adsorbent Carbopack C modified with 0.1% SP-1000 liquid. Temperature programmed from 70 to 225°C at 4°/min. Flow: 20 mL/min. Detector: FID. Reprinted with permission from the catalog of Supelco, Inc., Bellefonte, PA.

Molecular sieves as the Zeolites and Carbosieves™ separate molecules by size, as discussed in Chapter 3. They are best known for their ability to separate oxygen and nitrogen. A classic separation is shown in Figure 3.5.

Porous synthetic polymers of styrene and divinylbenzene have been made for GSC and find application in many separations, especially for samples containing water and other polar molecules (Figure 8.5). A new

1. Water
2. Methanol
3. Ethanol
4. Acetone
5. Methyl Ethyl Ketone
6. Tetrahydrofuran
7. p-Dioxane
8. Dimethyl Formamide

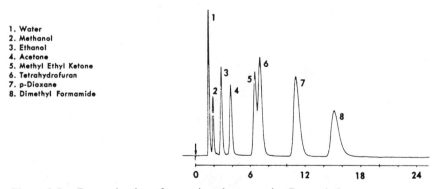

Figure 8.5. Determination of water in solvents, using Porapak Q porous polymer at 220°C. Courtesy of Varian Instruments.

series of these polymers can also separate oxygen and nitrogen.[7] Another popular porous polymer is called Tenax™; it is made from 2,6-diphenyl-p-phenylene oxide.

Open Tubular (OT) Columns. In their original and most simple form, OT columns contain a thin film of stationary liquid on their inside walls. Hence they are referred to as *wall coated open tubular*, or *WCOT*, columns. Compared to packed columns, they have low pressure drops and small amounts of stationary phase (and relatively large β). Originally they were made of stainless steel, and the early technology has been described by Ettre.[8] A revolution occurred in the late 1970s when the column material changed first to glass[9] and then to fused silica.[10] Because of the superior performance (inertness) and flexibility of fused silica columns, they have become the most popular type. Figure 8.6 is typical of their performance.

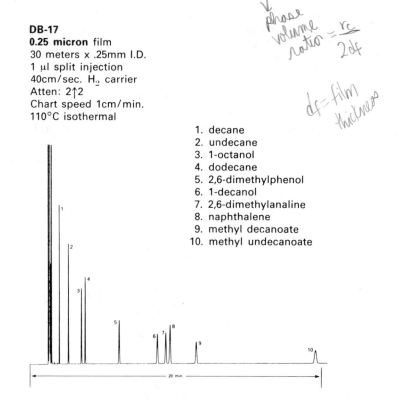

DB-17
0.25 micron film
30 meters x .25mm I.D.
1 μl split injection
40cm/sec. H₂ carrier
Atten: 2↑2
Chart speed 1cm/min.
110°C isothermal

1. decane
2. undecane
3. 1-octanol
4. dodecane
5. 2,6-dimethylphenol
6. 1-decanol
7. 2,6-dimethylanaline
8. naphthalene
9. methyl decanoate
10. methyl undecanoate

← 20 min →

Figure 8.6. Standard test mixture for OT columns run on a 30 m × 0.25 mm id (inside diameter) column of DB-17 with 0.25 μm film at 110°C. Courtesy of J & W Scientific.

A more severe test mixture, including acids and bases, is shown in Figure 8.7. Note the asymmetry of peak 6 and the TZ numbers for the members of the homologous series of esters—peaks 9, 11, and 12.

For a period of time, OT columns that had characteristics intermediate between those of WCOT and packed columns were popular. There were two types, but they were similar. Support coated open tubular (SCOT) columns had a thin layer of solid support coated on the inside wall of a capillary tube of larger diameter than that used for WCOT columns. This layer was coated with stationary liquid similar to packed columns. Porous layer open tubular (PLOT) columns were similar but made differently; for example, the solid support was added while the capillary tube was being drawn. With a few exceptions, SCOT and PLOT columns are no longer popular because wide diameter WCOT fused silica columns are as good, more stable (no layer to flake off), and easier to use. Subsequent discussion will be restricted to WCOT columns.

Coating the inside of a capillary tube requires pretreatment of the silica so that the liquid will wet the surface and stick to it.[11,12] Stable stationary phases have been attained recently by bonding the liquid to the silica and/ or crosslinking it, using a variety of methods including gamma irradiation.[13] The crosslinked bonded phase OT columns are very stable and can even be washed with (liquid) solvents for cleaning.

The thickness of the film of liquid, d_f, can be controlled, and a variety of thicknesses are commercially available. Thicker films can accommodate larger samples, but they are somewhat less efficient. When d_f is known, the phase volume ratio β can be calculated:

$$\beta = \frac{r_c}{2d_f} \tag{4}$$

where r_c is the radius of the OT column. Thus, for OT columns, the fundamental partition coefficient K can be calculated for a given analyte, since the partition ratio k is easily determined from the chromatogram obtained, and

$$K = k\beta \tag{5}$$

It is not easy to obtain partition coefficients for packed columns because the volume of the stationary phase is more difficult to calculate.

The *outside* of fused silica columns must be coated with a polyimide to extend their life and keep them flexible.

To increase the capacity of WCOT columns, wider diameters and heavier loadings are used. These columns have largely replaced the SCOT

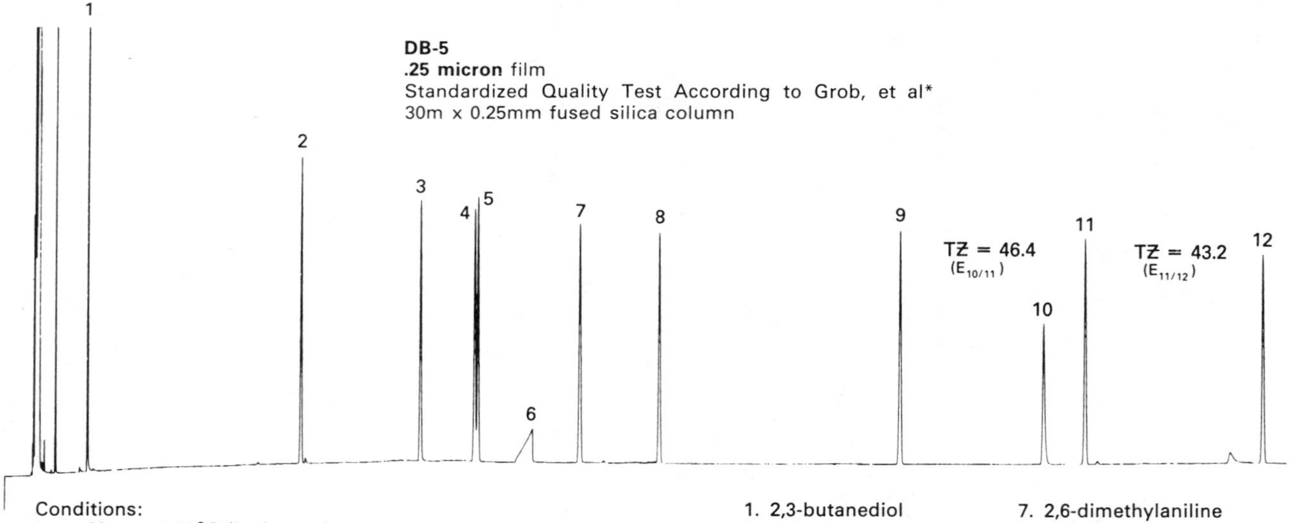

Figure 8.7. "Grob" test mixture run on 30 m DB-5 OT column. Courtesy of J & W Scientific.

columns and are known as *wide-bore* or *mega-bore* columns. The two most common sizes of WCOT columns are compared in Table 4, but several other sizes are commercially available, and an excellent discussion of the various types has been given by Duffy.[14] The wide-bore columns are best for low boiling mixtures and will accommodate larger samples; the conventional columns give the highest efficiencies and may permit the use of shorter lengths and thus shorter analysis times. Ettre[15] has summarized the effect of column diameter and liquid film thickness on OT column performance.

Comparison of Packed and OT Columns. Some of the important differences between packed and conventional OT columns are given in Table 5. OT column technology is advancing rapidly and OT columns are becoming more popular; wide-bore OT columns can be used when a column is needed with characteristics intermediate between the two extremes.

Detectors

Virtually every conceivable means of detecting gases and vapors has been exploited in designing GC detectors, and over one hundred have been described. The two most popular ones, the thermal conductivity detector (TCD) and the flame ionization detector (FID), will be described in some detail. They are classified (according to the criteria in Chapter 7) and compared in Table 6.

Thermal Conductivity Detector (TCD). The TCD cell is a metal block in which cavities have been drilled to accommodate the transducers, which can be either thermistors or resistance wires (so-called *hot wires*, Figure 8.8). Thermistors are most sensitive at low temperatures and find limited

TABLE 4   Comparison of Fused Silica WCOT Columns

	Conventional	Wide Bore
Outside diameter	0.40 mm	0.70 mm
Inside diameter	0.25 mm	0.53 mm
Film thickness (d_f)	0.25 μm	1–5 μm
Phase volume ratio (β)	250	130–25
Column length	15–60 m	15–30 m
Flow	1 mL/min	5 mL/min
H_{min}	0.3 mm	0.6 mm
n_{eff}	3000/m	1200/m
Typical sample size	50 ng	15 μg

TABLE 5   Comparison of Packed and WCOT Columns

	$\frac{1}{8}''$ Packed	WCOT
Outside diameter	3.2 mm	0.40 mm
Inside diameter	2.2 mm	0.25 mm
d_f	5 μm	0.25 μm
β	15–30	250
Column length	1–2 m	15–60 m
Flow	20 mL/min	1 mL/min
n_{tot}	4,000	180,000
n_{eff}	2,000/m	3,000/m
H_{min}	0.5 mm	0.3 mm
Advantages	Lower cost	Higher efficiency
	Easier to make	Faster
	Easier to use	More inert
	Larger samples	Fewer columns needed
	Better for fixed gases	Better for complex
		mixtures

use. The wire filaments can be supported on holders or be mounted concentrically in a cylindrical cavity. The latter arrangement permits the volume of the cavity to be minimized, which is highly desirable. A typical design, shown in Figure 8.9, has four cavities; two is the minimum—one each for the reference and the sample flows. A special low volume TC cell has been produced by Hewlett Packard; to keep the volume small, only one cavity is used and the two gas streams (reference and sample) are passed through it alternately.

The resistance wires, made of tungsten or a tungsten-rhenium alloy, are heated with a DC source to a temperature above the block temperature

TABLE 6   Comparison of TCD and FID

	TCD	FID
Classifications	Concentration	Mass flow rate
	Bulk property	Specific property
	Universal	Slightly Selective
	Nondestructive	Destructive
Characteristics		
Sensitivity	5,000 mV mL/mg	10^{-2} C/g
Detectivity (MDQ)	10^{-10} g/mL	10^{-12} g/sec
Minimum sample size	10^{-8} g	10^{-10} g
Linearity	10^4	10^6

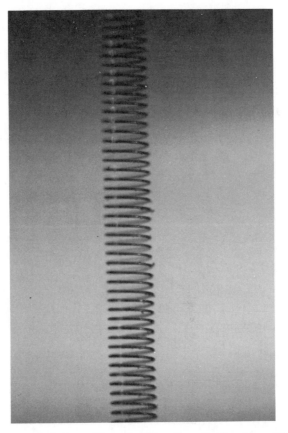

Figure 8.8.   Magnified view of hot wire filament coil. Courtesy of the Gow-Mac Instrument Co.

and lose heat to it at a rate dependent upon the thermal conductivity of the gas in the cavity. Thus, the temperature, and hence the resistance, of the hot wire depends upon the thermal conductivity of the gas in the cavity. The wires are incorporated into a Wheatstone bridge circuit (Figure 8.10) and produce a voltage imbalance when an analyte passes through one side of the cell. The wires can be heated at constant voltage or constant current or be maintained at a constant temperature by varying the current or voltage. Keeping the filament temperature constant requires a more complex circuit, and the output signal is derived from the electrical changes that are necessary to bring the bridge back to null, rather than from the direct voltage imbalance. Sensitivity is increased by heating the filament to a higher temperature with the power supply, and it is a function

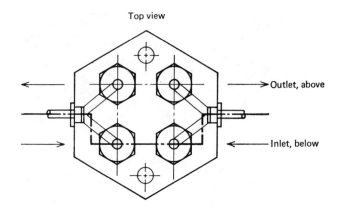

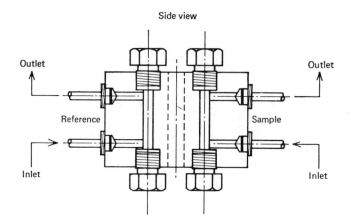

Figure 8.9. Typical TCD, designed for four concentric hot wires. Courtesy of the Gow-Mac Instrument Co.

of the difference in temperature (ΔT) between the filament and the block. If a TCD is operated in air, the filaments are quickly oxidized and burn out. Even small leaks in the chromatographic system will result in gradual destruction of the hot wires.

The carrier gas must have a thermal conductivity that is very different from the analytes to be detected, so that the most commonly used gases are helium or hydrogen. Some relative thermal conductivities are given in Table 7. With a high TC carrier gas like He, the filament runs relatively cool; when a sample enters the sample cavity, that filament gets hotter, its resistance goes up, and a signal is produced. Other mechanisms contribute to the loss of heat from the filament, and response values cannot

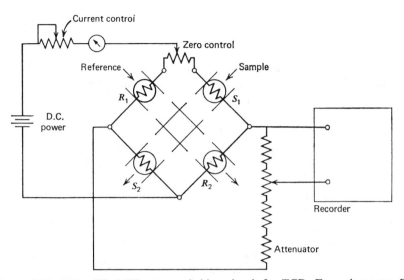

Figure 8.10. Simplified Wheatstone bridge circuit for TCD. Four elements: R_1 and R_2 for reference, and S_1 and S_2 for example.

TABLE 7   Thermal Conductivities and TCD Response Values for Selected Compounds

Compound	Thermal Conductivity[a]	RMR[b]
Carrier Gases		
Argon	12.5	—
Carbon dioxide	12.7	—
Helium	100.0	—
Hydrogen	128.0	—
Nitrogen	18.0	—
Samples		
Ethane	17.5	51
n-Butane	13.5	85
n-Nonane	10.8	177
i-Butane	14.0	82
Cyclohexane	10.1	114
Benzene	9.9	100
Acetone	9.6	86
Ethanol	12.7	72
Chloroform	6.0	108
Methyl iodide	4.6	96
Ethyl acetate	9.9	111

[a] Relative to He = 100.
[b] Relative molar response in helium. Standard: benzene = 100.

be calculated from thermal conductivities alone. For quantitative analysis, response values must be determined; they are included in Table 7. Remember that the TCD is a concentration detector and the peak areas it produces are flow dependent. Although the TCD is only moderately sensitive, it is a universal, simple, rugged, and inexpensive detector.

Flame Ionization Detector (FID). The FID is a small oxy–hydrogen flame in which the sample is burned, producing some ions in the process. These ions are then collected, constituting a small current, which is the signal. A typical FID design, Figure 8.11, shows the column effluent mixed with hydrogen and led to a small burner tip that is swept by a high flow of air for combustion. An igniter is necessary for remote lighting of the flame. The collector electrode is biased about 300 V relative to the flame tip, and the signal current is amplified. The exact mechanism of flame ionization is still not known, but the ionization efficiency, while very low, is sufficient to give a good sensitivity and linearity. The flow rates of hydrogen and air must be optimized for a particular detector design (and to

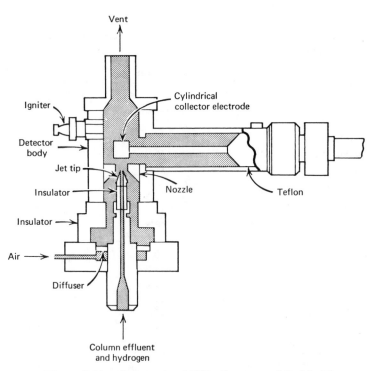

Figure 8.11.   Schematic of FID. Courtesy of Perkin-Elmer.

a lesser extent, analyte); typical values, as shown in Figures 8.12 and 8.13, are a hydrogen flow approximately equal to the column flow and an air flow of at least 300 mL/min. For OT columns, a make-up flow of carrier gas is necessary to bring the total up to about 30 mL/min, since OT flow rates are of the order of only 1 mL/min. As noted earlier, hydrogen is becoming increasingly popular as the carrier gas of choice for OT columns; this requires changes in gas flows and has prompted new FID designs.[16]

The FID is nearly universal, detecting all compounds with a hydrocarbon backbone, but not those listed in Table 8. Water is a compound that often produces badly tailed peaks on GC columns, so it may be advantageous that the FID does not detect it. However, the FID cannot be used for analysis of fixed gases. Like the TCD, response factors must be determined for good quantitative analysis.[17]

In summary, the FID is the detector of choice for organic analysis because of its sensitivity, which allows it to be used with OT columns. It has a good stability and linearity, but it requires additional gases for its operation.

Other Detectors. Most of the other commercially available detectors are highly selective and sensitive and find use in specialized analyses, including inorganic analyses. One exception is the gas density balance,[18] which is a universal type of detector and finds use in molecular weight determinations (see Chapter 6) and for the analysis of corrosive materials.

The electron capture detector (ECD) was invented by Lovelock in 1961 and is probably the third most used detector. As its name implies, it is selective for materials that capture electrons—halogen- and nitrogen-containing compounds such as pesticides and unsaturated compounds such as the polynuclear aromatics. It is an ionization detector, but unlike the FID it is a concentration type and a bulk property type detector. As such it is an exception to our generalization that bulk property detectors are not very sensitive.

The ECD has a high standing current caused by the ionization of the carrier gas by a radioactive source. It requires nitrogen or a mixture of argon plus 5% methane as a carrier gas and a radioactive source. Formerly tritium was used as the excitation source, but ^{63}Ni is more common now because it has a higher temperature limit of 350°C. The electrons caused by the ionization produce a large current output from the detector, but the presence of an electron-capturing analyte decreases this current as the electrons are absorbed. This absorption of electrons follows an equation similar to Beer's law.

The ECD has an MDQ detectivity of about 10^{-12} g/mL for ideal an-

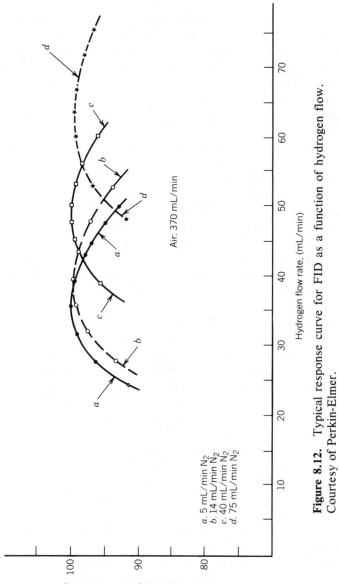

Figure 8.12. Typical response curve for FID as a function of hydrogen flow. Courtesy of Perkin-Elmer.

128

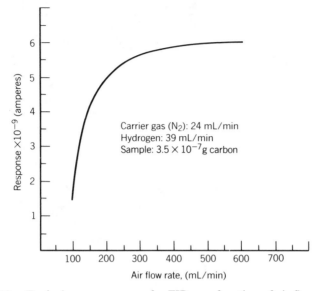

Carrier gas (N_2): 24 mL/min
Hydrogen: 39 mL/min
Sample: 3.5×10^{-7}g carbon

Figure 8.13. Typical response curve for FID as a function of air flow. Courtesy of Perkin-Elmer.

alytes and a linearity of three or four orders of magnitude. It is very easily contaminated and somewhat troublesome to operate.

The characteristics of some other detectors are summarized in Table 9. The popular MS and FTIR detectors (often used on-line) are not included because they are discussed in Chapter 11. A comparison of the working ranges of the most common detectors is shown in Figure 8.14.

For more details on the detectors see the books by David[19] and Dressler,[20] the review of multiple detection in GC,[21] and the section on detectors in Uden's review of inorganic analysis.[22]

TABLE 8 Some Compounds Not Detected by the Flame Ionization Detector

He	O_2	NO	CS_2
Ar	N_2	NO_2	COS
Kr	CO	N_2O	$SiCl_4$
Ne	CO_2	NH_3	$SiHCl_3$
Xe	H_2O	SO_2	SiF_4

TABLE 9 Other GC Detectors

Name	Operating Principle	Selective for[a]	Ref.
Ionization Type Detectors			
Thermionic ionization (TID) or nitrogen/phos (NPD) or alkali flame ionization	Alkali salt vapors cause chemical ionization; electric or flame heated	P, N, X	23
Photoionization (PID)	UV lamp causes photoionization	Aromatics	24
Helium ionization (HID)	He carrier gas ionized by tritium	Universal, but used for gases	25
Emission Type Detectors			
Flame photometric (FPD)	Flame excitation causes emission	S, P	26
Plasma emission	Plasma excitation causes emission	Metals	22
Other Detectors			
Hall electrolytic conductivity (HECD)	Catalytic reaction to form HX, H_2S, NH_3; measure conductance	S, N, X	27
Thermal energy analyzer (TEA)	Thermal fragmentation; chemiluminescence	Nitro- & nitroso-	28
Radioactivity (RAD)	β or γ detectors	Radioactive	29

[a] X = halogen.

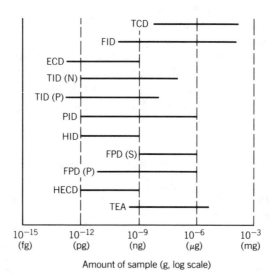

Figure 8.14. Comparison of typical working ranges for common GC detectors.

Instrumentation Summary

Although GC has become a mature analytical technique, improvements are still being made. New and better OT columns are becoming available so that their use has expanded, requiring new and better instrument designs. Multiple columns and column switching are becoming popular; a series of papers on the subject of *multidimensional GC* describes the current developments.[30]

A series of tests of instrument performance has been proposed[31] and can be used to determine: (1) packed column injector performance, (2) column oven heating and cooling specifications, and (3) FID performance ratings.

THE STATIONARY PHASE

The selection of a stationary phase is one of the most important decisions in setting up a method. Because GC is so easy to perform, the process of selection has often been made on a trial-and-error basis and hundreds of liquids have been employed. The theoretical approach, while very complex, has been successful in aiding in the choice of liquid and has permitted the reduction in number of necessary liquids. Refer to Chapter 3 for the background information on this topic.

The stationary liquid phase is coated on a solid support and packed into a column (packed column GC), or it is coated on the wall of an open tube (WCOT), as has been discussed. The higher efficiency of OT columns has reduced the necessity for many selective liquids, and the number of OT columns necessary to analyze for "all" types of analytes is smaller than for packed columns.

A persistent problem with liquid phases is their upper temperature limits. As GC is used at higher and higher temperatures in an attempt to extend its usefulness to higher and higher boiling analytes, the stationary phase vapor pressure gets higher and higher and is evidenced as *column bleed*. Thus, a major objective has been to find liquids with increasing boiling points—polymers of high molecular weight, special new polymers, and bonded phases. The latter were extensions of the reactions used to deactivate solid supports by silanization, mentioned earlier. The active hydroxyl groups on the surface of most solid supports can be reacted with a variety of chemicals to form ether or ester linkages resulting in the addition of alkyl chains bonded to the solid support. These bonded supports represent a special class of materials for use in GC,[32] but they have found even more extensive use in LC. Therefore, a more extensive discussion is given in Chapter 9. For OT columns, we have already discussed the special crosslinking reactions and surface preparations that are used to achieve stability of liquid phases.

Liquid phases also have lower temperature limits, represented by their melting points or glass transition temperatures. In most cases these temperatures are low enough so that they are well below the normal working GC temperatures. However, there are some exceptions, so the minimum should also be checked before use. Also, some special applications use subambient temperatures at the beginning of a programmed temperature run, and these are often below the recommended minimum temperature of the stationary phase.

Typical Liquids

Silicone polymers, differing in the extent to which they contain polar functional groups, have become one popular class of liquid phases. They are discussed as a group, followed by all the other miscellaneous types, including those with special selectivities. Finally, some recommendations are given for the selection of a minimum number of stationary phases.

The Silicone Polymers. The silicone polymers have the backbone structure shown in Figure 8.15 in which all the alkyl groups are shown as methyl. Replacement of the methyl groups with others of higher polarity yields polymers of increasing polarity. Since one company, Ohio Valley Speciality Chemical, produces a complete line of these products, Table 10 contains a list of their products by OV number. Those indicated by an asterisk are available in a special formulation on OT columns, and most of them contain 1% vinyl groups to improve stability. Other manufacturers' phases equivalent to the OV phases are given in Table 11.

Other Important Liquids. The classification of liquid phases has required the selection of a standard nonpolar liquid phase to which the others can be compared. The chemical chosen for this purpose because of its non-

Figure 8.15. Basic structural unit of silicone polymers used as stationary phases in GC.

polarity and its availability is the branched hydrocarbon squalane, $C_{30}H_{62}$. Because it has an upper temperature limit of only 125°C, a longer chain hydrocarbon called Apolane 87 has been synthesized. Its formula is $C_{87}H_{176}$, and it can be used up to 260°C, although it is slightly more polar than squalane.

Another relatively nonpolar liquid, but one with a very high temperature limit, is a carborane silicone polymer called Dexsil, first synthesized by Olin Chemical in 1964 and now manufactured by Analabs and by a company bearing the Dexsil name. Its structure is shown in Figure 8.16. Three polymers are available that differ in the substitution on the methyl silicone backbone in a way similar to the silicones discussed above: Dexsil 300 is dimethyl silicone, as shown in the figure; Dexsil 400 is the methyl–phenyl analog; and Dexsil 410 is the methyl–cyanopropyl analog. The temperature range over which they can be used is approximately 50 to 400°C.

A popular series of polar phases is the polyethyleneglycols; of this series the Carbowax™ line is one of those commercially available. The structure of this polymer is

$$OH—(CH_2—CH_2—O)_n—H$$

and the approximate average weight is given as a numerical value in the naming of the particular polymer. The highest boiling member available in this series is Carbowax 20M, which has an average molecular weight of 20,000. It can be used from 60 to 225°C in packed columns and up to 280°C in special OT bonded configurations.

Another popular polar phase is diethyleneglycol succinate, DEGS:

$$(—CH_2—CH_2—O—CH_2—CH_2—O—\overset{\|}{\underset{O}{C}}—CH_2—CH_2—\overset{\|}{\underset{O}{C}}—O—)_n$$

Its temperature range is 20 to 200°C, and it finds use for separating fatty acid methyl esters.

Several selective phases have size exclusion properties in addition to their solvent properties. A mixture of 5% of the clay Bentone 34 with 5% of a nonpolar liquid is one of the few phases that can separate the xylene isomers.[33] The class of compounds known as *liquid crystals* also finds special application in isomer separations.[34] When silver salts are incorporated into a liquid phase such as a glycol, special selectivity for olefins is obtained.[35] Optical isomers have been separated by GC, and this topic is included in Chapter 11.

TABLE 10 Characteristics of Silicone Polymers by OV-Number

Name	Type	Solvent	Temp. Min. (°C)	Temp. Limit (°C)	Viscosity	Average Mol Weight	ΔI McReynolds Constants[a]						
							①	②	③	④	⑤	⑥	⑦
OV-1*	Dimethylsilicone gum	Toluene	100	325–375	Gum	$>10^6$	16	55	44	65	42	32	23
OV-101	Dimethylsilicone	Toluene	20	325–375	1,500	3×10^4	17	57	45	67	43	33	23
OV-3	Phenylmethyldimethylsilicone, 10% phenyl	Acetone	20	325–375	500	2×10^4	44	86	81	124	88	55	46
OV-7	Phenylmethyldimethylsilicone, 20% phenyl	Acetone	20	325–375	500	1×10^4	69	113	111	171	128	77	66
OV-11	Phenylmethyldimethylsilicone, 35% phenyl	Acteone	0	325–375	500	7×10^3	102	142	145	219	178	103	92
OV-17*	Phenylmethylsilicone, 50% phenol	Acetone	20	350–375	1,300	4×10^3	119	158	162	243	202	112	105
OV-61	Diphenyldimethylsilicone	Acetone	20	325–375	>50,000	4×10^4	101	143	142	213	174	99	86
OV-73*	Diphenyldimethylsilicone gum	Toluene	20	325–350	Gum	8×10^5	40	86	76	114	85	57	39
OV-22	Phenylmethyldiphenylsilicone	Acetone	20	350–375	>50,000	8×10^3	160	188	191	283	253	133	132

Phase	Name	Solvent		Temp. range			①	②	③	④	⑤	⑥	⑦
OV-25	Phenylmethyldiphenylsilicone	Acetone	20	350–375	>100,000	1×10^4	178	204	208	305	280	144	147
OV-105	Cyanopropylmethyldimethylsilicone	Acetone	20	275–300	1,500		36	108	93	139	86	74	29
OV-202	Trifluoropropylmethylsilicone	Chloroform	0	250–275	500	1×10^4	146	238	358	468	310	206	56
OV-210	Trifluoropropylmethylsilicone	Chloroform	20	275–350	10,000	2×10^5	146	238	358	468	310	206	56
OV-215*	Trifluoropropylmethylsilicone gum	Chloroform	20	250–275	Gum		149	240	363	478	315	208	56
OV-225*	Cyanopropylmethylphenylmethylsilicone	Acetone	20	250–300	9,000	8×10^3	228	369	338	492	386	282	150
OV-275	Dicyanoallylsilicone	Acetone	20	250–275	20,000	5×10^3	629	872	763	1106	849	686	318
OV-330	Silicone carbowax copolymer	Acetone	30	250–275	500	5×10^3	222	391	273	417	368	284	158
OV-351	Polyglycolnitroterephthalic	Chloroform	50	250–270	Solid		335	552	382	583	540	—	—
OV-1701*	Dimethylphenylcyano substituted polymer	Acetone	20	300–325	Gum		67	170	153	228	171	—	—

ᵃ Key: ① Benzene; ② butanol; ③ 2-pentanone; ④ nitropropane; ⑤ pyridine; ⑥ 2-methyl-2-pentanol; ⑦ 2-octyne

135

TABLE 11   Equivalent Silicone Polymer Liquid Phases[a]

Ohio Valley Number	Other Designations					
OV-1, 101	SP-2100 RSL-150	SE-30	DB-1	SPB-1	SF-96	DC-200
———	SE-54	SE-52	DB-5	SPB-5	RSL-200	
OV-7				SPB-20		
OV-11				SPB-35		
OV-17	SP-2250		DB-17	RSL-300		
OV-210, 202, 215	SP-2401	QF-1	DB-210			
OV-225			DB-225	RSL-500		
OV-351	SP-1000	(FFAP)	AT-1000			
———	SP-2300	Silar 5C				
———	SP-2340	Silar 10C		CS-10		
———	SP-2100	UC W982				
OV-1701			DB-1701			

[a] Equivalent Rohrschneider/McReynolds Values.

Recommended Liquids for Packed Columns. Some consensus has been achieved in selecting the most popular liquid phases for packed column GC; Table 12 lists most of them. A committee headed by Hawkes[36] found that 6 liquids are most popular (see Table 12), followed by another 20 that are also common, and 13 that are used for special separations. More recently, Yancey[37] found 5 liquids to be the most popular; his list is nearly the same as Hawkes's, as can be seen in Table 12. For packed column operation, a selection of these 5 or 6 phases should provide a good "stable" of columns for most separations. The order of arrangement in Table 12 is one of increasing "polarity" according to Rohrschneider/McReynolds constants, which will be discussed next. The wide range of

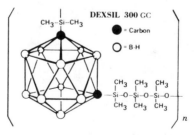

Figure 8.16.   Basic structural unit of Dexsil polymers.

TABLE 12 Rohrschneider/McReynolds Constants for Common Liquid Phases[a-c]

Liquid Phase	x'	y'	x'	u'	s'
Squalane	0	0	0	0	0
Apolane 87	21	10	3	12	25
OV-1	16	55	44	65	42
OV-101	17	57	45	67	43
OV-3	44	86	81	124	88
Dexsil 300	41	83	117	154	126
Dexsil 400	60	115	140	188	174
Dexsil 410	85	165	170	340	180
OV-17	119	158	162	243	202
Tricresyl phosphate	176	321	250	374	299
OV-202 and OV-210	146	238	358	468	310
OV-225	228	369	338	492	386
Carbowax-20M	322	536	368	572	510
DEGS	492	733	581	833	791
Silar 10C	523	757	659	942	801
1,2,3-Tris(2-Cyanoethoxy) propane (TCEP)	592	857	759	1031	917
OV-275	629	872	763	1106	849

[a] From Supina and Rose.[39]
[b] Recommended by Hawkes[36]: OV-101, OV-17, Carbowax $\geq$ 4000, OV-210, DEGS, and Silar 10C.
[c] Recommended by Yancey[37]: OV-101, OV-17, Carbowax 20M, OV-202, and OV-225.

numerical values indicates a wide range of polarities, and the most used columns cover most of the range.

Rohrschneider/McReynolds Constants

Chapter 3 ended by noting that an alternative way of estimating the *polarity* of a stationary phase was to use *probes*. Basically, the extent of interaction between the stationary phase and a given sample will be reflected in the adjusted retention time (or partition ratio or retention index, I) of the solute. Thus, by choosing as solute probes those chemicals that are thought to have particularly strong selective interactions, one can get a measure of the relative magnitude of that interaction from its retention index.

Rohrschneider[38] was the first to suggest a list of probes and a method of organizing the data. His original paper is in German, but Supina and Rose[39] have described it in English. The five solute probes are listed in Table 13. For the stationary liquid phases, squalane was chosen as the least polar and assigned Rohrschneider constants of zero.

TABLE 13   Compounds Used for Liquid-Phase Characterization

	Probes Used by	
Designation	Rohrschneider	McReynolds
a	Benzene	Benzene
b	Ethanol	n-Butanol
c	2-Butanone (MEK)	2-Pentanone
d	Nitromethane	Nitropropane
e	Pyridine	Pyridine
		2-Methyl-2-pentanol
		Iodobutane
		2-Octyne
		1,4-Dioxane
		cis-Hydrindane

To get the polarity constants for the other stationary phases, each solute in Table 13 was run on squalane and on the liquid phase of interest at 100°C and 20% liquid loading. The retention index I was determined for each analyte, and the difference between the two values on the two phases (ΔI) was obtained. The ΔI summed for all five probes is given as

$$\Delta I = ax + by + cz + du + es \qquad (6)$$

where $a, b, c, d,$ and e represent the five solutes and $x, y, z, u,$ and s represent the five Rohrschneider constants for that liquid phase. Thus a has a value of 100 for benzene and a value of zero for the other four solutes; b is 100 for ethanol and zero for the others, and so on. Each of the x, y, z values equals ΔI divided by 100, and the ΔI value in Eq. (6) is actually the sum of the five individual ΔI values.

In 1970 McReynolds[40] went one step further. He reasoned that ten probes would be better than five and that some of the original five should be replaced by higher homologs. His probes are also listed in Table 13. (Also, he did not divide his values by 100, so his constants are 100 times those of Rohrschneider; these larger values are the ones that have become popular). The first 5 of his 10 constants are the ones listed in Table 12, where their designations are distinguished by the use of a prime—x' instead of x. McReynolds published constants for over 200 liquid phases, and this procedure has been followed by others to provide Rohrschneider/McReynolds constants for all the liquid phases in use.

In fact, it has turned out that McReynolds was wrong—ten probes were not better than five. Therefore, most compilations of McReynolds constants list only five, six, or seven values. In fact, later attempts to expand on this method have tried to reduce the number of constants to four, as we will see.

Rohrschneider went on to show that one can use the liquid phase constants in reverse to get values that would represent the *polarities* of other solutes. The values in Table 14 were compiled by him by running a new solute on five liquid phases whose Rohrschneider constants had been previously determined. The five ΔI values thus obtained are substituted into five equations [like Eq. (6)], which are solved simultaneously for the five unknowns a, b, c, d, and e. As expected, the a value for toluene is close to 100 (108), the b value for propanol is close to 100 (105 for *n*-propanol), and the c value for acetone is nearly 100 (95) because these three solutes have the same functional groups as the original probes. To answer the question "How polar is chloroform," one can look at the five constants and tell what functional groups (or interactions) it is "like" and what groups it is not like. In principle, one can determine the solute constants for the components of any sample, multiply them by the Rohrschneider/McReynolds constants for a variety of common columns, sum the values, and thus find the best column for the separation. This approach is not the one analysts have taken to select the best liquid phase for a given separation. Before we take a look at the procedures that are used, let us note some related uses that can be made of the Rohrschneider/McReynolds constants.

We have already seen a list of recommended liquid phases (Table 12),

TABLE 14   Rohrschneider Constants for Some Solutes[a]

Solute	a	b	c	d	e
Benzene	100	0	0	0	0
Ethanol	0	100	0	0	0
2-Pentanone	0	0	100	0	0
Nitromethane	0	0	0	100	0
Pyridine	0	0	0	0	100
i-Propanol	− 18.2	+ 95.9	+ 15.8	− 6.5	+ 2.1
n-Propanol	− 9.4	+ 105.3	+ 0.3	+ 6.6	− 7.5
Acetone	− 5.3	− 4.6	+ 94.9	+ 7.9	+ 5.6
n-Butyl acetate	− 3.8	− 13.3	+ 57.3	+ 13.9	+ 20.0
Propionaldehyde	+ 13.3	− 1.0	+ 74.9	+ 4.8	+ 1.3
Di-*n*-butyl ether	+ 17.3	+ 9.8	+ 29.7	− 12.5	− 2.8
Cyclohexane	+ 32.1	− 22.5	− 21.6	+ 4.1	+ 29.7
Chloroform	+ 69.7	+ 28.9	− 72.6	+ 53.1	− 6.3
Toluene	+ 108.3	+ 3.8	+ 8.8	− 7.0	− 7.6

[a] From Rohrschneider (38).

which are given in order of increasing polarity. That these phases are different can be seen directly from the five McReynolds constants listed for each. While the five values will not tell you which liquid phase is best for a given separation, any one value can indicate a particularly strong interaction. For example, tricresyl phosphate has an unusually high y' value, indicating that it interacts strongly with alcohols and probably has a strong tendency to hydrogen-bond with hydrogen-bonding donors.

Another example is OV-202 and OV-210. The identical constants show that these two polymers are identical and differ only in chain length and viscosity. This type of comparison was important in the early days of GC when the supplies of some of the polymers used as liquid phases became exhausted. New polymers were made and shown to be equivalent to the old ones; one example is OV-210, which replaced QF-1.

Finally, the sum of the first five McReynolds values has been used to compare the polarities of silicone polymers on OT columns. It has already been stated that the polarity of the silicone polymers can be increased by increasing the percentage of phenyl groups in the polymer. Figure 8.17 shows a plot for five polymers on bonded fused silica WCOT columns (except for SP-2250, which is from packed column data). The increasing McReynolds values clearly show the validity of this method of specifying stationary phase polarity.

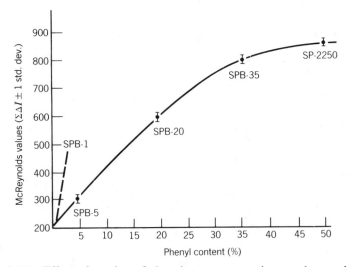

Figure 8.17. Effect of number of phenyl groups on stationary phase polarity as measured by McReynolds values. Reprinted from the *Supelco Reporter*, Vol. IV, No. 3, May 1985 with the permission of Supelco Inc., Bellefonte, PA.

Methods for Selecting Stationary Phases

There are a variety of ways to approach the important task of choosing a stationary phase (a column) for a GC analysis. Chapter 12 elaborates on the searching of the literature and logical decision-making processes based on the information that is at hand. The procedures outlined here are more concerned with the fundamentals of matching a liquid phase to a separation problem, drawing on theoretical principles as much as possible.

The need for reliable methods for selecting liquid phases is greatly diminished when highly efficient OT columns are used. Indeed, it is often the case that a 15-m column of DB-5 (or the equivalent nonpolar bonded phase) will separate most of an analyst's samples. Nevertheless, the following methods have contributed to our understanding of the chromatographic process and are worthy of consideration.

Quasi-Theoretical Approach. Several groups of workers have taken the McReynolds constants and similar data and attempted to combine them with our present knowledge of intermolecular attractions as presented in Chapter 3. In a pair of papers, Hartkopf[41] suggested that four probes (not five or ten) would be sufficient to characterize a liquid phase. He proposed to measure the following:

Dispersion interactions with benzene.

Dipole orientations with nitroethane.

Hydrogen bonding with two probes: dioxane as an acceptor of a hydrogen bond, and either *n*-propanol or chloroform as a hydrogen-bond donor.

His proposal has not attracted much attention, probably because the data have not been determined, while the McReynolds data are available.

The McReynolds data were standardized and subjected to principal component analysis by several groups of workers who were able to reduce the data to three statistical components. Burns and Hawkes[42] further refined the calculations to produce four quasi-theoretical indices that measure dispersion, polarity, acidity, and basicity. Hawkes has described this process in a more recent paper[43] in which his group confirmed and refined these calculations with spectroscopic measurements. In addition to justifying their approach, they provide four indices for each of the 26 common liquid phases that were identified earlier as being the most important.[36] The dispersion index is calculated from refractive indices, but the other three indices are based at least partially on chromatographic data.

Other developments will undoubtedly follow these interesting studies, which have already achieved the goal of providing numerical estimations of the types of intermolecular forces. However, a set of four numbers is still not very useful to an analyst trying to choose a liquid phase, and another approach called the *window method* provides an alternative.

Window Method of Purnell. In 1975 Laub and Purnell[44] suggested a largely empirical method for finding the optimum *mixture* of liquids that would give the best separation for a given sample containing at least three analytes. Their only assumption was a common one—that GC retentions as measured by partition coefficients are additive in proportion to the volume fractions of the liquids used to make the mixed stationary phase; that is,

$$K_{\text{mix}} = \phi_X K_X + \phi_Y K_Y \tag{7}$$

where X and Y are the two liquids chosen to be used in a mixture as the stationary phase and ϕ is the volume fraction of each. Rearranging Eq. (7), we get an equation for a straight line:

$$K_{\text{mix}} = K_Y + \phi_X(K_X - K_Y) \tag{8}$$

Figure 8.18a shows the lines obtained for each of four hypothetical cases.

If we let $(K_X - K_Y)$ equal ΔK, and we now consider two analytes, A and B, whose relative retention α is

$$\alpha = \frac{K_A}{K_B} \tag{9}$$

we can write

$$\phi_X = \frac{(K_{(X)A} - \alpha K_{(X)B})}{(\alpha \Delta K_B - \Delta K_A)} \tag{10}$$

A plot of ϕ_X versus α for the hypothetical four-component mixture yields a diagram like that shown in Figure 8.18b, if α is always defined to keep its value greater than 1. Laub and Purnell called this type of diagram a *window diagram*, and that term has persisted. The best mixture of X and Y will be the one that provides the largest α value; in Figure 18b that point occurs at a ϕ_X of 0.12, predicting that a mixture of 12% X and 88% Y will give the best resolution of the six possible pairs of analytes. The predicted α of 1.23 is quite good.

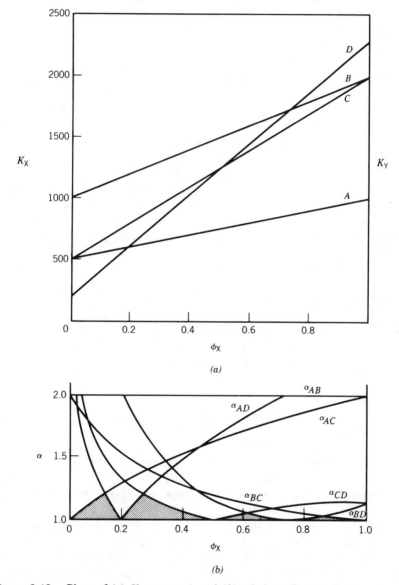

Figure 8.18. Plots of (*a*) K_R versus ϕ and (*b*) window diagram for four hypothetical analytes. Reprinted from ref. 44 with permission.

To recap, first one needs to determine the partition coefficients for each of the analytes on each of the two liquid phases—each chosen because it is effective in separating most of the sample analytes when used alone. Second, one plots these K values versus ϕ_X by using the two K values for each analyte at $\phi_X = 0$ and $\phi_X = 1.0$ and joining these two points with a straight line. Then, the window diagram is drawn using ϕ_X values calculated from Eq. (10) at various α values.

The best separation will occur at the ϕ_X value where the highest α occurs. If two or more windows have approximately the same α value, the best one will be the smallest one since it will usually provide the shortest analysis time. The length of column needed to get a given resolution can be calculated with the equation for n_{req} presented in Chapter 4 and the assumption of a reasonable plate height.

The desired liquid phase mixture can be prepared by combining the proper ratio of the individual phases and packing it into a column. These two individual packings will be available from the early part of the experiment where the individual partition coefficients were measured. Finally, the whole calculation can be done by computer, as described by Purnell and co-workers.[45]

TEMPERATURE EFFECTS

Temperature is one of the two most important variables in GC. Retention times decrease as temperature increases because partition coefficients are temperature dependent in accordance with the Clausius–Clapeyron equation:

$$\log p^0 = -\frac{\Delta \mathcal{H}}{2.3 \mathcal{R} T} + \text{constant} \qquad (11)$$

where $\Delta \mathcal{H}$ is the enthalpy of vaporization and is assumed to be constant over the range of temperatures investigated. The analyte's vapor pressure p^0 decreases with increasing temperature, resulting in a decrease in partition coefficient and in retention volume.

Verification of this relationship in GC is provided by Figure 8.19, in which the log of the net retention volume is plotted versus $1/T$ in accordance with Eq. (11). The slope of each line is proportional to that analyte's enthalpy of vaporization, and the fact that straight lines are obtained indicates that the enthalpy is constant as assumed.

To a first approximation, the lines in Figure 8.19 are parallel, indicating that the enthalpies of vaporization for these compounds are nearly the

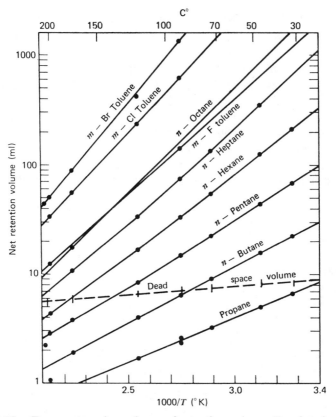

Figure 8.19. Temperature dependence of retention volume. Reprinted with permission from W. E. Harris and H. W. Habgood, *Talanta* **1964**, *11*, 115. Copyright 1964, Pergamon Journals, Ltd.

same. A closer inspection reveals that the lines diverge slightly at low temperatures. From this observation we can draw the generalization that GC separations are usually better at lower temperatures. But look at the two analytes, *n*-octane and *m*-fluorotoluene; their lines cross. At a temperature of about 140°C they cannot be separated; at a lower temperature the toluene elutes first; at a higher temperature the reverse is true. It is not common for elution orders to reverse, but it can happen!

The effect of temperature on column efficiency is quite complex,[46] and no generalizations can be drawn. Usually it is minor and of considerably less importance than the effect on column thermodynamics. The latter is of such importance that temperature programming has become very important.

Programmed Temperature GC (PTGC)

PTGC is the process of increasing the temperature of the column oven during a run. As we have just seen, the increasing temperature will cause the partition coefficients of the analytes still on the column to decrease, and they will move faster through the column, yielding decreased retention times. The effect can be seen in Figure 8.20.

Some major differences between the two runs can be seen and are typical of PTGC. For a homologous series, the retention times are logarithmic under isothermal conditions, as we saw in Chapter 6, and they are linear when programmed. The programmed run was begun at a lower temperature (50°C) than the one used for the isothermal run (150°C), which

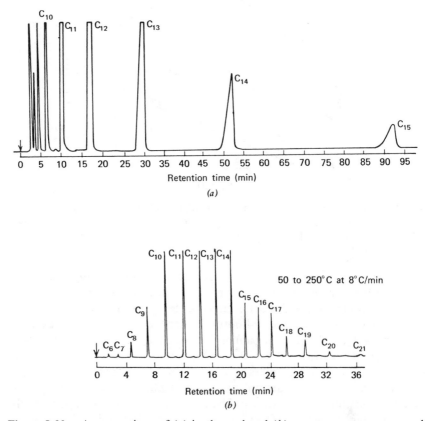

Figure 8.20. A comparison of (*a*) isothermal and (*b*) temperature-programmed separation of *n*-paraffins. Figure courtesy of McNair and Bonnelli, *Basic Gas Chromatography*, Varian, 1968.

facilitated the separation of the low-boiling paraffins. It ended at a higher temperature (250°C), which increased the number of paraffins detected. The peak widths are about equal in PTGC, while some fronting is evidenced in the higher boilers in the isothermal run. Since the peak widths do not increase in PTGC, the heights of the late-eluting analytes will be increased (peak areas are constant), providing better detectivity. The advantages and disadvantages of PTGC are summarized in Table 15.

Programmed temperature operation is good for screening new samples. A maximum amount of information about the sample composition is obtained in minimum time. Usually one can tell when the entire sample has been eluted, which is often a difficult judgment to make with isothermal operation.

The theory of PTGC has been thoroughly treated by Harris and Habgood[47] and by Mikkelsen.[48] The following discussion has been taken from a simple but adequate treatment by Giddings.[49]

The dependence of retention volume on temperature was illustrated in Figure 8.19. Let us determine approximately what temperature increase is necessary to cut a given retention volume in half; that is:

$$\frac{K_2}{K_1} = \frac{1}{2} \tag{12}$$

or

$$\frac{K_1}{K_2} = 2 = \frac{\exp(-\Delta \mathcal{H}/\mathcal{R}T_1)}{\exp(-\Delta \mathcal{H}/\mathcal{R}T_2)} = \exp\left[\frac{\Delta \mathcal{H}}{\mathcal{R}\overline{T}}\left(\frac{\Delta T}{\overline{T}}\right)\right] \tag{13}$$

where ΔT is the difference between the two temperatures T_1 and T_2, and $\overline{T}$ is the average of the two temperatures. Taking the log and rearranging,

TABLE 15   Advantages and Disadvantages of PTGC

Advantages	Disadvantages
1. Better separation for wide boiling mixtures	1. Has additional instrument requirements
2. Constant peak width and shape	2. Ghost peaks may occur
3. Decreases time of analysis	3. Time required for cooldown
4. Sample introduction less critical	4. Fewer stationary phases can be used
5. No loss of quantitative accuracy	5. Subject to baseline drift and noise due to bleeding
6. Lower limits of detection	

we get

$$\Delta T = \frac{0.693 \Re T^2}{\Delta \mathcal{H}} \tag{14}$$

Assuming Trouton's rule that $\Delta \mathcal{H}/T_b = 21$ and a boiling temperature of 227°C (500 K) for a typical sample,

$$\Delta T = \frac{0.693 \times 2 \times (500)^2}{21(500)} \approx 30°C \tag{15}$$

As an approximation, then, an increase in temperature of 30°C will cut the retention volume in half. (This rule of thumb is also useful for iso-thermal operation, of course).

The effect of temperature programming on the migration of a typical analyte through a column is shown in Figure 8.21, where the 30°C value is used to generate the step function. The retention ratio R_R will double for every 30° since the rate of movement of the analyte through the column is equal to ($R_R \times u$). Final elution from the column is arbitrarily taken as occurring at 265°C, as shown in the figure. In actuality, the movement of the analyte through the column would proceed by the smooth curve also shown in the figure since the temperature programming would be gradual and not stepwise, as assumed by our model.

If l is taken as the distance the analyte moved through the column in the last 30° increment, then one-half is the distance it moved in the previous (next-to-last) 30°, one-quarter in the 30° before that, and so on. The sum of these fractions approaches $2l$, which must equal the total length L of the column. Hence the analyte traveled through the last half of the column in the last 30°; it started moving slowly and speeded up as the temperature increased. The operation of PTGC can be envisioned as follows: the sample is injected onto the end of the column, and its components largely remain there; as the temperature increases, the analytes "boil off" and move down the column at increasing rates until they elute. It is for these reasons that the injection technique is not critical in PTGC and that all peaks have about the same peak widths, since they spend about the same amount of time actively partitioning down the column.

For a variety of reasons, isothermal operation is often preferred in the workplace. If an initial screening is done by PTGC, one might wish to know what isothermal temperature would be the single best one to use. Giddings has called this isothermal temperature the *significant temperature*, T'. Using reasoning based on the 30°C value, he has found that

$$T' = T_f - 45 \tag{16}$$

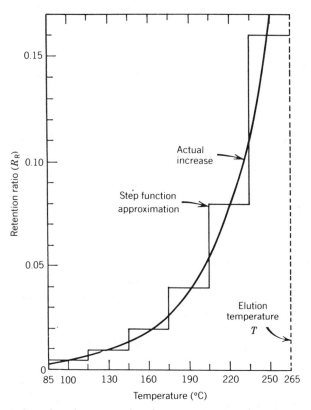

Figure 8.21. Step-function approximation to programmed-temperature GC. Reprinted from ref. 49 courtesy of the Chemical Education Division, American Chemical Society.

where T_f is the final temperature, the temperature at which the analyte(s) of interest eluted in the PT run.

Two other important variables are the programming rate and the column length. In general, one does not vary the length but uses a rather short column (and lower temperatures). The rate is often chosen to be fast enough to save time but slow enough to get adequate separations, which results in rates between 5 and 10°/min. However, for OT columns, there is considerable evidence that slow rates around 2.5°/min are preferable.[50]

Some chromatographs are provided with ovens that can be operated below ambient temperature, thus extending the range of temperature programming. Examples can be found in the extensive review of cryogenic GC by Brettell and Grob.[51]

SPECIAL TOPICS

Inorganic GC

Most inorganic compounds are not volatile enough for analysis by GC, with the exception of some fixed gases like CO_2 and SO_2. Consequently inorganic GC is usually treated as a separate topic that is concerned with the formation of volatile derivatives and special detectors and elemental analysis. A thorough review has been written by Uden,[22] who has also contributed a chapter to a recent book on inorganic chromatography.[52] A comprehensive review by Bachmann[53] includes a large section on inorganic GC. General discussion about derivatization is included in Chapter 11.

Inverse GC

As the name implies, this technique is used to produce data that are the opposite or inverse of normal GC methods. Its objective is to get information about large nonvolatile molecules rather than the usual small ones. Consequently the large molecules are used as stationary phases and are then subjected to investigation with small molecules that serve as probes. Further details can be found in the review by Gilbert.[54]

Flow Programming

Increasing the flow during a run is an alternative to PTGC. Although it has some advantages, the technique has not become very popular, but it has been thoroughly reviewed.[55,56]

Preparative Scale GC

GC is not ideally suited to large sample sizes and cannot be scaled up in size without major changes in the columns used. While it was once popular, it is no longer so, most prep work being done by LC. Some laboratories still maintain prep equipment, and at least one line is still commercially available. Details can be found in references 57–59.

Other Special Techniques

When GC replaced distillation as a method of analyzing petroleum products, the GC equivalent of distillation was formulated by groups like ASTM. Appropriately enough, the methods are called *simulated distillation*.[60]

TABLE 16   Evaluation of GC

Advantages	Disadvantages
1. Efficient, selective, and widely applicable	1. Samples have to be volatile
2. Fast	2. Not suitable for thermally labile samples
3. Inexpensive and simple	3. Fairly difficult for large samples (preparative)
4. Easily quantitated	4. Theory inadequate, so some trial and error may be required
5. Requires only a small sample	
6. Nondestructive	

Several special sampling techniques should be mentioned. Many analyses are simplified by taking a sample from the vapor over the liquid sample. This procedure is known as *headspace* analysis.[61] Another method known as *purge-and-trap* is used to concentrate a sample to improve detection limits. It is very popular in pesticide residue and drinking water analyses.

Pyrolysis GC for analyzing nonvolatiles including polymers was discussed in Chapter 6, and *derivatization* is in Chapter 11.

EVALUATION

Table 16 concludes this chapter with a summary of the advantages and disadvantages of GC.

REFERENCES

1. S. Kenworthy, J. Miller, and D. E. Martire, *J. Chem. Educ.* **1963,** *40,* 541.

2. M. J. E. Golay in *Gas Chromatography,* V. J. Coates, H. J. Noebels, and I. S. Fagerson (eds.), Academic Press, New York, 1958, pp. 1–13.

3. F. L. Bayer, *J. Chromatogr. Sci.* **1986,** *24,* 549.

4. J. F. Schneider, S. Bourne, and A. S. Boparsi, *J. Chromatogr. Sci.* **1984,** *22,* 203.

5. C. A. Cramers and J. A. Rijks, *Adv. Chromatogr. N. Y.* **1979,** *17,* 101.

6. D. M. Ottenstein, *J. Chromatogr. Sci.* **1973,** *11,* 136.

7. G. E. Pollock, D. O'Hara, and O. L. Hollis, *J. Chromatogr. Sci.* **1984,** *22,* 343.

8. L. S. Ettre, *Open Tubular Columns,* Plenum, New York, 1965.

9. W. Jennings, *Gas Chromatography with Glass Capillary Columns*, 2d ed., Academic Press, New York, 1980.

10. M. L. Lee, F. J. Yang, and K. D. Bartle, *Open Tubular Column Gas Chromatography; Theory and Practice*, Wiley, New York, 1984.

11. See, for example, M. L. Lee, R. C. Kong, C. L. Woolley, and J. S. Bradshaw, *J. Chromatogr. Sci.* **1984**, *22*, 136.

12. R. R. Freeman (ed.), *High Resolution Gas Chromatography*, 2d ed., Hewlett-Packard, Avondale, Pa., 1981.

13. J. A. Hubball, P. R. Di Mauro, E. F. Barry, E. A. Lyons, and W. A. George, *J. Chromatogr. Sci.* **1984**, *22*, 185.

14. M. L. Duffy, *Am. Lab.* **1985**, *17*(10), 94.

15. L. S. Ettre, *Chromatographia* **1984**, *18*, 477.

16. R. K. Simon, Jr., *J. Chromatogr. Sci.* **1985**, *23*, 313.

17. H. Y. Tong and F. W. Karasek, *Anal. Chem.* **1984**, *56*, 2124.

18. J. T. Walsh and D. M. Rosie, *J. Gas Chromatogr.* **1967**, *5*, 232.

19. D. J. David, *Gas Chromatographic Detectors*, Wiley-Interscience, New York, 1974.

20. M. Dressler, *Selective Gas Chromatographic Detectors*, Elsevier, Amsterdam, 1986.

21. I. S. Krull, M. E. Swartz, and J. N. Driscoll, *Adv. Chromatogr. N. Y.* **1984**, *24*, 247.

22. P. C. Uden, *J. Chromatogr.* **1984**, *313*, 3–31.

23. B. Kolb and J. Bischoff, *J. Chromatogr. Sci.*, **1974**, *12*, 625. P. L. Patterson, *Chromatographia* **1982**, *16*, 107. For a recent application see C. M. White, A. Robbat, Jr., and R. M. Hoes, *Anal. Chem.* **1984**, *56*, 232.

24. J. N. Driscoll, J. Ford, L. Jaramillo, J. H. Becker, G. Hewitt, J. K. Marshall, and F. Onishak, *Am. Lab.* **1978**, *10*(5), 137; P. Verner, *J. Chromatogr.* **1984**, *300*, 249.

25. R. S. Brazell and R. A. Todd, *J. Chromatogr.* **1984**, *302*, 257.

26. M. C. Bowman and M. Beroza, *Anal. Chem.* **1968**, *40*, 1448.

27. R. J. Anderson and R. C. Hall, *Am. Lab.* **1980**, *12*(2), 108.

28. J. H. Phillips, R. J. Coraor, and S. R. Prescott, *Anal. Chem.* 1983, *55*, 889.

29. P. A. Rodriguez, C. R. Culbertson, and C. L. Eddy, *J. Chromatogr.* **1983**, *264*, 393.

30. See pp. 2–34 in *J. Chromatogr. Sci.* **1986**, *24*.

31. J. Q. Walker, S. F. Spencer, and S. M. Sonchik, *J. Chromatogr. Sci.* **1985**, *23*, 555.

32. J. N. Driscoll and I. S. Krull, *Am. Lab.* **1983**, *15*(5), 42. R. K. Gilpin, *J. Chromatogr. Sci.* **1984**, *22*, 371.

33. S. F. Spencer, *Anal. Chem.* **1963**, *35*, 592.

34. H. Keller and E. von Schivizhoffen, *Adv. Chromatogr. N. Y.* **1968**, *6*, 247.

35. J. H. Purnell in *Gas Chromatography—1966*, A. B. Littlewood (ed.), Institute Petroleum, London, 1967, p. 3.

36. S. Hawkes, D. Grossman, A. Hartkopf, T. Isenhour, J. Leary, J. Parcher, S. Wold, and J. Lancey, *J. Chromatogr. Sci.* **1975**, *13*, 115.

37. J. A. Yancey, *J. Chromatogr. Sci.* **1986**, *24*, 117.

38. L. Rohrschneider, *J. Chromatogr.* **1966**, *22*, 6.

39. W. R. Supina and L. P. Rose, *J. Chromatogr. Sci.* **1970**, *8*, 214.

40. W. O. McReynolds, *J. Chromatogr. Sci.* **1970**, *8*, 685.

41. A. Hartkopf, *J. Chromatogr. Sci.* **1974**, *12*, 113; A. Hartkopf, S. Grunfeld, and R. Delumyea, *J. Chromatogr. Sci.* **1974**, *12*, 119.

42. W. Burns and S. J. Hawkes, *J. Chromatogr. Sci.* **1977**, *15*, 185.

43. E. Chong, B. deBriceno, G. Miller, and S. Hawkes, *Chromatographia* **1985**, *20*, 293.

44. R. J. Laub and J. H. Purnell, *J. Chromatogr.* **1975**, *112*, 71.

45. R. J. Laub, J. H. Purnell, and P. S. Williams, *J. Chromatogr.* **1977**, *134*, 249.

46. W. E. Harris and H. W. Habgood, *Talanta* **1964**, *11*, 115.

47. W. E. Harris and H. W. Habgood, *Programmed Temperature Gas Chromatography*, Wiley, New York, 1966.

48. L. Mikkelsen, *Adv. Chromatogr. N. Y.* **1966**, *2*, 337.

49. J. C. Giddings, *J. Chem. Educ.* **1962**, *39*, 569.

50. L. A. Jones, S. L. Kirby, C. L. Garganta, T. M. Gerig, and J. D. Mulik, *Anal. Chem.* **1983**, *55*, 1354.

51. T. A. Brettell and R. L. Grob, *Am. Lab.* **1985**, *17*(10), 19; and (11), 50.

52. P. C. Uden in *Inorganic Chromatographic Analysis*, vol. 78 of Chemical Analysis Series, J. C. MacDonald (ed.), Wiley, New York, 1985, chapter 5.

53. K. Bachmann, *Talanta*, **1982**, *29*, 1.

54. S. G. Gilbert, *Adv. Chromatogr. N. Y.* **1984**, *23*, 199.

55. R. P. W. Scott in *Advances in Analytical Chemistry and Instrumentation*, Vol. 6, C. N. Reilly (ed.), Wiley, New York, 1968, p. 271.

56. L. S. Ettre, L. Mazor, and J. Takacs, *Adv. Chromatogr. N. Y.* **1969**, *8*, 271.

57. G. W. A. Rijnders, *Adv. Chromatogr. N. Y.*, **1966**, *3*, 215.

58. D. T. Sawyer and G. L. Hargrove in *Advances in Analytical Chemistry and Instrumentation*, Vol. 6, C. N. Reilly (ed.), Wiley, New York, 1968, p. 325.

59. A. Zlatkis and V. Pretorius, *Preparative Gas Chromatography*, Wiley, New York, 1971.

60. L. G. Chorn, *J. Chromatogr. Sci.*, **1984**, *22*, 17.

61. H. Hachenbert and A. P. Smith, *Gas Chromatographic Headspace Analysis*, Heyden, London, 1977.

The original form of chromatography introduced by Tswett at the turn of the century was LC carried out in columns, but its development was slow until after GC was introduced. Then, the principles and theories of GC were applied to LC, and it too developed quickly. From the late 1960s until the present, the pace of research in LC has accelerated, and many further improvements and modifications are to be expected in the future. This is especially true in the biomedical area since LC is much more amenable to the analysis of large biomolecules than is GC.

The range of operating modes is much greater in LC than in GC because both phases, stationary and mobile, affect the separation and because a wide range of stationary phases can be used in LC. Stationary phases include many types of bonded phases with a wide range of polarities, and also materials with ion exchange and sieving (size exclusion) properties.

Operating conditions vary widely, from low pressure to high pressure, short columns to long columns, narrow columns to *infinite diameter* columns, analytical samples to prep samples, and so on.

This monograph cannot do justice to all these types of LC. For the most part the discussion will focus on *HPLC*, high performance (and high pressure) LC performed on analytical samples using the most common types of columns and conditions. Brief mention and typical references will be given to other types whenever possible. We will begin with an overview of the different operating modes.

CLASSIFICATION OF LC METHODS

The original classification of LC methods in Chapter 1 included five types: liquid–solid (LSC), liquid–liquid (LLC), bonded phase (BPC), ion exchange (IEC), and size exclusion (SEC). The many other types of LC, such as *ion chromatography* and *ion-pair chromatography*, are discussed at the appropriate places within these five main types.

Bonded phases have largely replaced the liquid-coated columns used in LLC and have become the most important type of stationary phase used in LC. The data in Table 1, taken from a recent users' survey,[1] show that the most used form of LC is *reversed* phase using a bonded support. In total, over two-thirds of the work is done with bonded phases. Further,

TABLE 1   User's Survey of LC Usage[a]

	Preference (%)
Mode Most Often Used	
Bonded, reverse phase	56.1
C_{18} (ODS)	38.5
C_8	11.4
C_4	1.0
C_1	1.2
Phenyl-	4.0
Bonded, normal phase	12.6
LSC	12.4
IEC	9.8
SEC	6.7
Particle Size (μm)	
30	1.0
20–25	1.7
10–15	38
5–7	53
3–4	6.1
Particle Shape	
Spherical	83
Irregular	17

[a] Taken from ref. 1 with permission.

the survey reports a trend toward smaller particle sizes; the most popular range is now 5–7 μm. And, finally, it found that spherical supports are used preferentially by a ratio of about 5:1.

According to our earlier classification, the stationary phase can be a solid, a liquid, or a bonded phase. In the latter two cases, the phase must be coated on, or bonded to, particles of a porous solid support. Only a few materials have found widespread use as stationary solid supports; they are silica, synthetic polymers such as the styrene–divinylbenzene copolymer, diatomaceous earths, and some polysaccharides. The most common types and uses are given in Table 2.

Silica is the most widely used support. In LSC and SEC it is used by itself; in LLC it is coated with a liquid phase, and in BPC and IEC its surface is reacted with a modifier to give it special surface characteristics. Above a pH of about 8 it is sufficiently soluble in aqueous solutions that significant column degradation results with extended usage, thus limiting its use to the lower pHs.

TABLE 2   Characteristics of Stationary Phases

Chemical Composition	Used for	Limitations
Silica	LSC, LLC, BPC, IEC, SEC	Soluble at pH $\geqslant 8$
Styrene–divinylbenzene	IEC, BPC	
Polysaccharides	SEC, IEC	Compressible
Other polymers	SEC	
Diatomaceous earths	LLC	

The styrene–divinylbenzene copolymer structure is shown in Figure 9.1. The divinylbenzene causes crosslinking in the polymer and provides structural rigidity. The extent of crosslinking is controlled by the percentage of divinylbenzene used in the polymerization reaction. Typical values range from 4 to 16% and are usually indicated in the naming of the resin; for example, $\times 4$ and $\times 8$. The greater the crosslinking, the greater the rigidity, brittleness, and selectivity, and the less the swelling and permeability. For use in IEC, ionic functional groups are added to

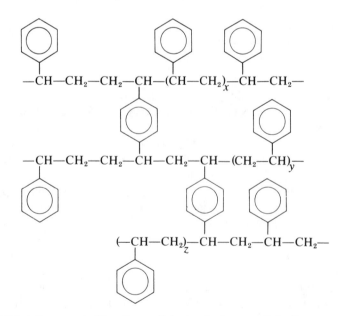

Figure 9.1.   Representation of crosslinked polystyrene–divinylbenzene polymer structure.

the resin, and for BPC nonionic groups are attached. All these stationary phases will be described further in their respective sections.

Other characteristics common to the stationary phases are particle size, pore size, surface area, and shape. Although it would seem that better columns could be packed using spherical particles, there is no overwhelming evidence that that is true. As noted in Table 1, spherical particles are more popular, as are the smaller particle sizes.

Pore sizes can vary from 4 to 200 nm and form the basis for separations by SEC. In the other modes of LC, the pores must be large enough to admit the analytes to the interior of the resin. Thus, it has been found[2] that large pores ($\geq$25 nm) are necessary for large molecules, but that small pores (10 nm) give more selectivity for small molecules and at decreased partition ratios.

The trend toward smaller porous particles is a result of the theoretical prediction that efficiency should increase as particle diameter decreases. However, as the diameter is decreased and the efficiency increased, the pressure drop also increases, leading to the use of shorter columns. The net effect is that the shorter small-particle columns have about the same efficiency as the conventional ones.

An alternative method for achieving the effect of a small diameter is to coat a thin layer of porous solid on a solid core (such as glass). Such materials were widely used in the early days of LC and are usually called *pellicular* supports. Typically the porous layer is 1–2 μm thick, and the solid core has a diameter of about 40 μm. Pellicular solids are easier to pack than microporous solids, but they are less stable, have smaller capacities, and are more expensive. Since good packing methods are now known for the microporous solids, the pellicular solids have become much less popular. They are commonly used in guard columns (discussed later in this chapter).

Surface area is also important. The larger the surface area, the greater the possibilities for adsorption and the likelihood of increased retention times. Consequently stationary phases with small surface areas are preferred for short analyses, but large surface areas are preferred for difficult separations.

Over the years a variety of methods for packing columns has been tried. To produce high efficiency columns with small particles, the best method uses a wet slurry packing at high pressure. Originally a balanced-density solvent was recommended[3] to keep the particles suspended, but this required the use of expensive and toxic liquids such as tetrabromoethane. More recent work has demonstrated that good columns can be packed using common solvents like acetone or methanol. For many laboratories, the time required to learn how to slurry pack cannot be justified for the number of columns needed, and commercial columns are purchased. However, the procedures are well documented, and the necessary equipment is available, so labs can pack their own columns.

The size of conventional columns is 250 mm long × 4.6 mm (i.d.), and straight stainless steel tubes are usually used, although heavy wall glass is also possible. Some workers have gotten better performance with steel columns that have a polished inner surface, but a recent study[4] disputes this conclusion. More information about columns is included in the instrumentation section of this chapter.

Liquid–Solid Chromatography (LSC)

The original work by Tswett was carried out with a solid stationary phase (LSC), and most of the early LC work used silica gel as the stationary phase. For this reason it has become the convention to label LSC methods using silica gel as *normal phase* LC (NPLC). To generalize, a *normal* LC system is one that has a polar stationary phase and a nonpolar mobile phase. The opposite situation is called *reverse phase* LC (RPLC). We noted in Table 1 that RPLC is now the most common type of LC, and consequently these two terms, NPLC and RPLC, do not reflect current usage and can be misleading. Unfortunately they are well established in the chromatographic literature and therefore will be used in this monograph as just defined.

The mode of action in LSC is adsorption, but the process is quite complicated because molecules of the mobile phase compete with analyte molecules for the active sites on the solid surface and silica is energetically heterogeneous. Any water present in the system will be strongly attracted to the silica surface, and there is evidence that there can be two or three layers of water adsorbed on silica. The most strongly adsorbed water layer cannot be removed with dry solvents, but the other layers can be. To get silica completely dry requires heating to temperatures above 200°C. Because of its importance in LSC, silica has been thoroughly studied; further details can be found in a number of published works.[5,6]

Snyder[7] has thoroughly studied adsorption chromatography, and some of his results were summarized in Chapter 3. He recommends covering 50 to 100% of the stationary solid surface with a monolayer of water. This requires up to 0.04 g of water per 100 m² of surface, or about 4 to 15% water added to the stationary phase. This process is not as easy as it sounds, and a long time is required for the system to come to equilibrium. This necessity to control the solid surface activity coupled with the pH limitation mentioned earlier has contributed to the decreased use of LSC for analytical separations although it is still popular for preparative LC.

The mobile phase in LSC is chosen for the following reasons: (1) proper strength or polarity, (2) low viscosity, (3) compatibility with detector, and (4) volatility if the analytes are to be recovered by evaporation of the mobile phase. If silica gel is the stationary phase, the main component of the mobile phase should be nonpolar. Typical solvents are listed in Table 3 in increasing order of Snyder solvent parameter ϵ^0, so the ones

TABLE 3   Solvent Properties of Some Liquids

Solvent	Estimated ϵ^0 (Silica)	δ	η	Refractive Index
Fluoroalkanes	−0.19	~5.5	—	1.25
n-Pentane	0.00	7.1	0.23	1.358
Isooctane	+0.01	7.0	0.50	1.404
Cyclohexane	+0.03	8.2	1.00	1.427
Cyclopentane	+0.04	8.1	0.47	1.406
1-Pentene	+0.06	—	0.18	1.371
Carbon disulfide	+0.11	10.0	0.37	1.626
Carbon tetrachloride	+0.14	8.6	0.97	1.466
Xylene	+0.20	8.9	0.6–0.8	~1.5
Isopropyl ether	+0.22	~7.3	0.37	1.368
Isopropyl chloride	+0.22	~8.4	0.33	1.378
Toluene	+0.22	8.9	0.59	1.496
Chlorobenzene	+0.23	9.5	0.80	1.525
Benzene	+0.25	9.2	0.65	1.501
Ethyl ether	+0.29	7.4	0.23	1.353
Chloroform	+.031	9.3	0.57	1.443
Methylene chloride	+0.32	9.7	0.44	1.424
Methyl isobutyl ketone	+0.33	—	0.58	1.394
Tetrahydrofuran (THF)	+0.35	9.1	—	1.408
Ethylene dichloride	+0.38	9.7	0.79	1.445
2-Butanone (MEK)	+0.39	9.3	0.44	1.381
Acetone	+0.43	9.9	0.32	1.359
Dioxane	+0.43	10.0	1.54	1.422
Ethyl acetate	+0.45	9.6	0.45	1.370
Methyl acetate	+0.46	9.2	0.37	1.362
Amyl alcohol	+0.47	9.8	4.1	1.410
Dimethyl sulfoxide (DMSO)	+0.48	12.8	2.24	1.479
Nitromethane	+0.49	12.6	0.67	1.394
Acetonitrile	+0.50	11.7	0.37	1.344
Pyridine	+0.55	10.7	0.94	1.510
Propanol, n and i	+0.63	11.5	2.3	1.38
Ethanol	+0.68	12.7	1.20	1.361
Methanol	+0.73	14.4	0.60	1.329
Ethylene glycol	+0.86	14.7	19.9	1.427
Acetic acid	Large	—	1.26	1.372
Water	Large	21	1.00	1.333

near the top of the table are the ones to use. Also included in the table are the Hildebrand solubility parameter δ, the viscosity η, and the refractive index.

Choosing the optimum mobile phase polarity is the most difficult task in LSC. Often a mixture of liquids is necessary to get the best results. A simple approach uses the Snyder solvent parameter, as shown in Figures 9.2 and 9.3. In Figure 9.2, Yost and Conlon[8] show how the Snyder parameter varies with composition for mixtures of liquids in heptane. Theoretically, mixtures that have the same ϵ^0 value should produce equal retention volumes. A test of this concept is shown in Figure 9.3, where acrylamide was run on a silica gel column with the four mixtures indicated, each having an ϵ^0 value of 0.53 achieved by mixing a second solvent with chloroform. The only surprise is the chromatogram obtained with the mixture of ethanol and chloroform; it produced undesirable peak broadening for unknown reasons.

Another, and more complete, listing of ϵ^0 values for mixed solvents is shown in Figure 9.4, taken from the work of Saunders.[9] The dashed line at a value of 0.3 is given as an example of the use of these data. This line cuts through six different solvent mixtures, all of which should produce the same results for a given sample.

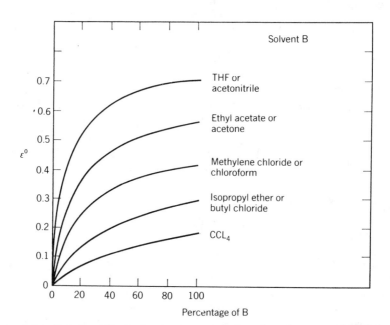

Figure 9.2. Polarity (ϵ^0) of mixed solvents as a function of composition. Solvent A is heptane. Courtesy of Perkin-Elmer.

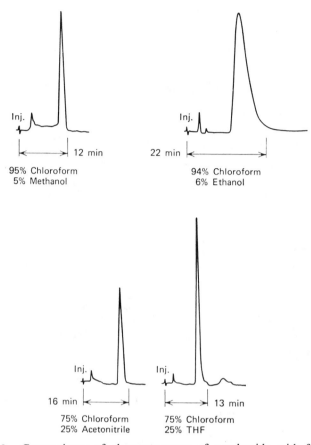

Figure 9.3. Comparisons of chromatograms of acrylamide with four mobile phases, each with a ϵ^0 value of 0.53. Column: 50 cm $\times$ 3 mm Sil-X. Flow: 1 mL/ min. Courtesy of Perkin-Elmer.

Using a similar procedure, or just trial and error, a mobile phase can be selected. The optimum situation exists when the analytes in the sample elute with partition ratios in the range of about 1 to 5 (Chapter 4). One rule of thumb[9] is that an increase of 0.05 units in ϵ^0 will decrease the partition ratio by one-half to one-quarter.

The term *isocratic* is used to describe LC runs in which the composition of the mobile phase is not varied during the run. If the sample has a wider range of partition ratios than can be accommodated isocratically, the composition of the mobile phase can be changed during the run. Such operation is called *gradient elution*. It is an important operating mode and is the subject of a monograph of that title.[10]

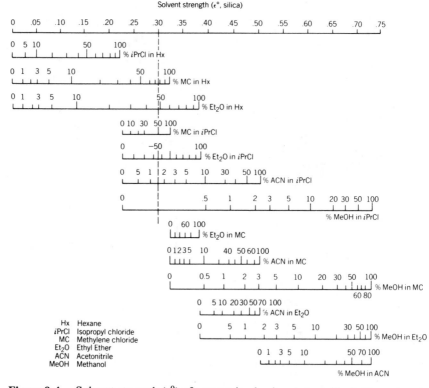

Figure 9.4. Solvent strength (ϵ^0) of some mixed solvents (on silica). Reproduced from the *Journal of Chromatographic Science* by permission of Preston Publications, Inc.

LSC, as we have described it, is NPLC, so the mobile phase should get more "polar" during a gradient run. In the simplest case, increasing amounts of a polar solvent are added to a nonpolar solvent such as pentane. The addition can be stepwise or continuous, the latter being preferred but requiring more extensive control. Some recommended gradients are given in Tables 4 and 5. Snyder and Saunders,[11] in their thorough discussion, have recommended that an optimum solvent program should have a change in ϵ^0 of 0.04 unit per column volume of solvent.

The chromatogram shown in Figure 9.5 uses the Series 2 gradient from Table 5 and represents the separation of an extreme mixture ranging from the nonpolar hydrocarbon squalane to the polar sugar glucose. Note, however, the irregular peak shapes, some of which can probably be attributed to displacement effects that can arise in drastic stepwise gradient runs.

Even greater versatility can be achieved if three solvents are mixed,

TABLE 4   Three Eluotropic Series for Silica Columns[a–c]

ϵ^0	Eluotropic Series		
	1	2	3
0.00	Pentane	Pentane	Pentane
0.05	4.2% PrCl in pentane	3% CH$_2$Cl$_2$ in pentane	4% Benzene in pentane
0.10	10% PrCl in pentane	7% CH$_2$Cl$_2$ in pentane	11% Benzene in pentane
0.15	21% PrCl in pentane	14% CH$_2$Cl$_2$ in pentane	26% Benzene in pentane
0.20	4% Ether in pentane	26% CH$_2$Cl$_2$ in pentane	4% EtOAc in pentane
0.25	11% Ether in pentane	50% CH$_2$Cl$_2$ in pentane	11% EtOAc in pentane
0.30	23% Ether in pentane	82% CH$_2$Cl$_2$ in pentane	23% EtOAc in pentane
0.35	56% Ether in pentane	3% Acetonitrile in benzene	58% EtOAc in pentane
0.40	2% Methanol in ether	11% Acetonitrile in benzene	
0.45	4% Methanol in ether	31% Acetonitrile in benzene	
0.50	8% Methanol in ether	Acetonitrile	
0.55	20% Methanol in ether		
0.60	50% Methanol in ether		

[a] After Snyder, *Modern Practice of Liquid Chromatography*, Kirkland (ed.), Wiley-Interscience, New York, 1971.
[b] All percentages are by volume.
[c] Abbreviations: PrCl, isopropyl chloride; EtOAc, ethyl acetate.

but the theory and the instrumentation are more complex. Several methods for selecting ternary mixtures will be described later.

The advantage of a gradient run is similar to that achieved using programmed temperature in GC in that it allows one to adjust the partition ratios of the analytes to get the best separations. The big disadvantage is the time required to reequilibrate the LC column; it is not as simple as cooling a GC oven. Table 5 included a recommended list of solvents to be used to recondition the column back to its original (nonpolar) status. About 10 column volumes of each solvent are required.

Many separations formerly done by LSC are now done with polar bonded phases, and these are discussed in a subsequent section. The main exceptions are preparative LC and TLC.

Liquid–Liquid Chromatography (LLC)

LLC originated with the Nobel Prize winning work of Martin and Synge in 1941. The separations are derived from the partitioning of analytes between two liquids, much as in liquid–liquid extraction, except that one liquid is held immobile on a stationary solid support. Huber and co-workers[12] have found good agreement between partition coefficients determined by chromatography and by static equilibrium measurements.

Since both phases in LLC are liquids, they must be immiscible in each

TABLE 5   Solvents for Incremental Gradient Elution

Series 1	Series 2
1. n-Heptane	1. n-Heptane
2. Carbon tetrachloride	2. Carbon tetrachloride
3. Heptyl chloride	3. Chloroform
4. Trichloroethane	4. Ethylene dichloride
5. n-Butyl acetate	5. 2-Nitropropane
6. n-Propyl acetate	6. Nitromethane
7. Ethyl acetate	7. Propyl acetate
8. Methyl acetate	8. Methyl acetate
9. Ethyl methyl ketone	9. Acetone
10. Acetone	10. Ethanol
11. n-Propanol	11. Methanol
12. Isopropanol	12. Water
13. Ethanol	
14. Methanol	
15. Water	

(items 4–7 of Series 2 braced as "Mixtures")

Column-Reconditioning Solvents

1. Ethanol
2. Acetone
3. Ethyl acetate
4. Trichloroethane
5. Heptane

[a] From R. P. W. Scott and P. Kucera, *J. Chromatogr. Sci.*, *1973*, **11**, 83, by permission of Preston Publications, Inc.
[b] From R. P. W. Scott and P. Kucera, *Anal, Chem. 1973*, **45**, 749.

other just as in extraction. This requirement mandates that one phase will be polar and the other nonpolar. In principle, either one could be the stationary phase, but the necessity to immobilize it on a solid support results most often in the stationary phase being the polar one, NPLC. Most of the liquid pairs listed in Table 6 represent NPLC. Also included are some ternary liquid systems in which two immiscible liquid mixtures are formed from the same three liquids by using the proper proportions.

LLC requires that the mobile phase be presaturated with the stationary phase, and that the column temperature be carefully controlled. It is relatively difficult to keep the stationary phase from washing off the column, and the number of liquid pairs that can be used is rather limited. For these reasons, bonded stationary phases that do not have these limitations have largely replaced conventional LLC.

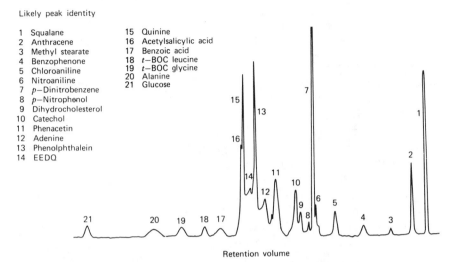

Likely peak identity

1	Squalane	15	Quinine
2	Anthracene	16	Acetylsalicylic acid
3	Methyl stearate	17	Benzoic acid
4	Benzophenone	18	*t*-BOC leucine
5	Chloroaniline	19	*t*-BOC glycine
6	Nitroaniline	20	Alanine
7	*p*-Dinitrobenzene	21	Glucose
8	*p*-Nitrophenol		
9	Dihydrocholesterol		
10	Catechol		
11	Phenacetin		
12	Adenine		
13	Phenolphthalein		
14	EEDQ		

Retention volume

Figure 9.5. LSC gradient elution chromatogram. Column: 50 cm × 5 mm i.d. Bio-Sil A. Flow: 0.5 mL/min. Sample: 10 mg in 50 μL. Reprinted with permission from R. P. W. Scott and P. Kucera, *Anal. Chem.* **1973**, *45*, 749. Copyright 1973, American Chemical Society.

TABLE 6   Some Liquid Phases for LLC

Stationary	Mobile
Binary	
1. Polyethylene glycol, such as Carbowax 400 or triethylene glycol	1. Hydrocarbon such as pentane, isooctane, or cyclopentane; or a hydrocarbon plus a small amount of more polar liquid such as chloroform
2. β,β'-Oxydipropionitrile	2. Same as 1
3. Dimethyl sulfoxide	3. Isooctane or hexane
4. Ethylenediamine	4. Same as 3
5. Tri-*n*-octylamine	5. Aqueous acid
6. Water–alcohol or glycol	6. Hexane–carbon tetrachloride
7. Hydrocarbons and polymers	7. Water–alcohol
Ternary	
1. Water–ethanol–isooctane	
2. Chloroform–cyclohexane–nitromethane	

Bonded Phase Chromatography (BPC)

The original bonded phases were made for GC since it was anticipated that chemical bonding would prevent the stationary phase from bleeding at high temperatures. The technology for reacting organic moities with solid surfaces was already established because of the silylation reactions used to deactivate GC solid supports (see Chapter 8). The original GC bonded phases are susceptible to decomposition from trace amounts of water or oxygen[13] and are therefore not widely used. Newer GC bonded phases are in use on OT columns as described in Chapter 8, and they are very stable, efficient, and popular. Similar bonded phases have been found to be excellent for LC. For a historical review, see the paper by Gilpin.[14]

At the beginning of this chapter it was noted that bonded phases have become the most popular form of LC, particularly those used in RPLC. The following discussion of the chemical reactions involved in producing bonded phases will be limited to these nonpolar stationary phases.

Most bonded phases are formed by reacting chlorosilanes with silica that has reactive silanol groups on its surface. (Other possibilities will be discussed later.) Figure 9.6 depicts some probable functional groups on the silica surface, and Figure 9.7 shows some typical deactivating reactions (or silanizations). For reverse phase supports, the R group, sometimes called a *ligate*, is an alkyl chain that can be up to 18 carbons long. If the reacting silane has only one reactive chloro group, the reaction is simple and replaces the active hydrogen with a dimethyloctadecylsilyl group. Bonded phases produced with this reaction are preferred for theoretical study because they are well-defined and regular; they are called *monofunctional phases.*

If the silane has more than one reactive chloro group, the reaction can be more complex, as shown. Some crosslinking can occur, resulting in an undefined polymer on the silica surface. The reaction shown in Figure 9.7 is only one of several possibilities.

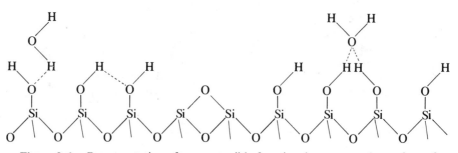

Figure 9.6. Representation of some possible functional groups on the surface of silica.

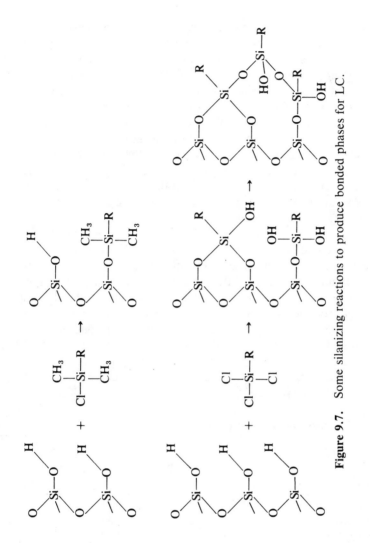

Figure 9.7. Some silanizing reactions to produce bonded phases for LC.

Unreacted chloro groups can be hydrolyzed to hydroxyl groups and then be reacted again with trimethylchlorosilane to eliminate as many hydroxyl groups as possible. This last silanization step is referred to as *end capping*; it also removes most of the previously unreacted, residual silanol groups on the surface of the silica where the larger ODS was not able to penetrate. Supports that have been end capped are usually found to have different selectivities from those that are not as fully reacted.

A variety of measurements, such as total carbon analysis and BET surface area, have been used to determine the effectiveness of the silylation reactions and the coverage of the silica surface. From these measurements the percentage of carbon on the surface can be calculated and is used to designate the degree of surface coverage. Values ranging from 2 to 30% carbon have been reported.

More recently, pyrolysis GC, ESCA, and FTIR have been used to characterize the surface of bonded layers. The reason for the intense interest is the fact that there are significant differences between bonded phases manufactured by different companies, and these investigators hope to find out why. One recent study[15] analyzed the decomposition products produced by the reaction of bonded reverse phases with HF, and it was able to determine the type of reaction (monofunctional or polyfunctional), the extent of end capping, and the distribution of lengths of alkyl groups. Some of the results of the study are summarized in Table 7. Such results help explain the differences between bonded phases manufactured by different companies.

TABLE 7   Analysis of Commercial ODS Phases[a]

| | | Amount of Liqate (g/g silica) | |
Manufacturer (particle size)	Type[b]	ODS	End Capping[c]
Du Pont (12 μm)	M	6.19	ND
LiChrosorb (10 μm)	D	5.50	ND
Nucleosil (10 μm)	T	—	3.24
Alltech (5 μm)	T	—	8.91
Beckman Ultrasphere IP	M	4.15	2.01
Adsorbo-sphere HS (7 μm)	M	8.51	2.05
Vydak 201-HS	M	6.71	4.62
Versapack (10 μm)	M	3.84	11.82
Techsil (5 μm)	T	—	2.38

[a] Adapted from S. D. Fazio et al., *Anal. Chem. 1985*, **57**, 1559 with permission. Copyright 1985, American Chemical Society.
[b] Type: M = monochloro; D = dichloro; T = trichloro.
[c] Trimethyl.

The wide use of these bonded phases has also stimulated a large effort to explain their surface structure and how they work. The so-called *solvophobic theory* of RPLC was elaborated originally by Horvath and Melander.[16] Their model assumes that the nonpolar bonded phase acts more like a solid than a liquid and attracts analytes by adsorption. The binding of an analyte to the surface reduces the surface area of the analyte exposed to the mobile phase, and it can be considered to be sorbed partially because of this solvent effect; that is, the analyte is sorbed because it is solvophobic. Sorption increases as the surface tension of the mobile phase increases.

Locke[17] concluded that bonded phases acted more like modified solids than thin liquid films, but that they were sufficiently different from solids to require a different theoretical treatment. Instrumental investigations of the bonded phase have been conducted by NMR, IR, and fluorescence and are included in the review by Gilpin[14] and in his more recent report.[18]

Bonded phases of other polarities can be prepared by substituting other groups for the octadecyl ligate. Some of the most common ones are listed in Table 8. The octyl ligate is also popular for RPLC, but very short chains like the dimethyl (C_2) ligate show significant polarity. The ligates in the table for NPLC are polar in accordance with our definition.

Since these ligates are bonded to the surface of the silica and will not wash off, it is no longer necessary to choose a mobile phase with a polarity opposite to that of the bonded phase. That is to say, a reverse phase support like ODS could be used with a relatively nonpolar organic mobile phase, thus establishing a system that does not fit into either category— reverse phase or normal phase. Some workers have given such a system the name *nonaqueous reverse phase*, or NARP.

In order for the analytes to penetrate into the bonded phase, the mobile phase used must "wet" the surface of the bonded phase. In RPLC, for example, silica that is thoroughly covered with ODS is quite hydrophobic

TABLE 8   Common Ligates for Bonded Supports Attached to Silica through a Silyl Linkage (See Text)

RPLC	NPLC
Octadecyl ($C_{18}H_{37}$, ODS)	Amino ($C_3H_6NH_2$)
Octyl (C_8H_{17})	Amino ($C_3H_6N(CH_3)_2$)
Dimethyl (C_2H_6)	Diamino ($C_3H_6NCH_2H_4NH_2$)
Phenyl	Cyano (C_2H_4CN)
Methoxy	Glycophase ($C_3H_6OCH_2CHCH_2OH$)
	$\qquad\qquad\qquad\qquad\quad |$
	$\qquad\qquad\qquad\qquad\;$ OH

and may not be wetted by a totally aqueous mobile phase, thus preventing the analytes from contacting the stationary phase. The addition of an organic modifier like methanol or acetonitrile to the mobile phase will usually solve this problem.

The mobile phases commonly used with the nonpolar bonded supports (RPLC) are mixtures of water and methanol or water and acetonitrile, often buffered. Millimolar concentrations of salts sometimes improve peak shapes. The requirement for high purity solvents extends to water, of course, and special high purity grades are available commercially. Separations are often better with the acetonitrile mixture, but methanol is less toxic and more easily disposed of. The expected order of elution would be from polar (hydrophylic) to less polar (hydrophobic). Further guidelines for choosing a mobile phase mixture are given later in this chapter. Remember, for gradient elution in RPLC, the mobile phase should get *less* polar; hence, the proportion of the organic solvent is increased during the run.

The nitrile and amine bonded phases are most popular for normal phase operation. Nonpolar liquids like the hydrocarbons and chlorocarbons are used as the mobile phases, although as noted above, even alcohols and aqueous mixtures can be used. The nitrile phase produces separations somewhat similar to plain silica in LSC.

Bonded phases have many advantages, as evidenced by their popularity. Compared to LSC, they come to equilibrium much faster, they do not show irreversible sorption or tailing, they can be used with a wide variety of solvents and trace amounts of water are not likely to be critical, and they come in a wide range of polarities. The most important disadvantage is that silica-based bonded phases must be used within the pH range of 2 to 7.5. Also, some care must be taken to prevent the polar functional groups like amine and nitrile from reacting with sample components and/or being oxidized.

The pH limitation has stimulated the production of bonded phases with the C_{18} ligates attached to a polymer rather than to silica. These supports, which have the C_{18} group on the phenyl ring of polystyrene–divinylbenzene, can be used in the pH range from 0 to 14. The polymer is highly crosslinked to give it the rigidity needed for high pressure operation. It does not seem to exhibit the interfering π interactions common to the polymer alone. Preliminary data indicate that the polymer-based materials are similar to those that are silica based.

Another type of bonded phase is intended to be used with biopolymers. In addition to the larger pore size that is required, these materials are intended to be used under milder conditions that will not denature the samples. When only water is used as the mobile phase, without any organic modifier, proteins will not be denatured and their hydrophobic characteristics will enable them to be separated on a nonpolar bonded phase. Many nonpolar bonded phases are too hydrophobic to be used with water alone, so new bonded phases have been made to accommodate these requirements. This type of chromatography is being called *hydrophobic interaction chromatography (HIC)*.[19] Supports are prepared by bonding C_{18}, C_8, C_5, or phenyl ligates on Sepharose or on a silica that has been previously bonded with a polyamide coating. Since the hydrophobic interaction between bioanalytes and the stationary phase is increased at high ionic strengths, gradient elutions can be achieved with a *decreasing* ionic strength gradient (the opposite of the effect in ion exchange, as we will see).

Ion Exchange Chromatography (IEC)

The name *ion exchange* aptly describes the process used chromatographically to separate ions and some polar molecules. Because ion formation is favored in aqueous solutions, the mobile phase in IEC is aqueous, usually buffered to a particular pH. Ionic exchange sites are immobilized on the stationary phase, represented in the following equation as R^-.

$$R^-A^+ + B^+ = R^-B^+ + A^+ \qquad (1)$$

The B^+ ion represents the analyte being separated from other cations, such as C^+ and D^+. The cations will be separated from each other if the resin (stationary phase) has a selective affinity for the various analyte cations. The cation A^+, which was the cation already on the resin, must not be too strongly held by the resin or exchange will not occur. It should be a component of the mobile phase, and its concentration in the mobile phase can be used to control the partition ratios of the analyte ions according to the equilibrium in Eq. (1).

The previous example was an exchange of cations and is known as *cation exchange chromatography*. If, on the other hand, the resin contains cationic sites, it is capable of exchanging anions, and the process is known as *anion exchange chromatography*. In both cases there may be some adsorption of analytes on the resin itself, thus complicating the mechanism, but these secondary effects will be ignored in this discussion.

Classical IEC. Ion exchange resins have been available for a long time as rather large beads used primarily at low pressures for rather crude separations. As early as 1951 Moore and Stein[20] published their classic work showing an IE separation of amino acids. One of their separations is shown in Figure 9.8; it is a gradient analysis, and it requires a long time for completion. Instruments designed specifically for such analyses were

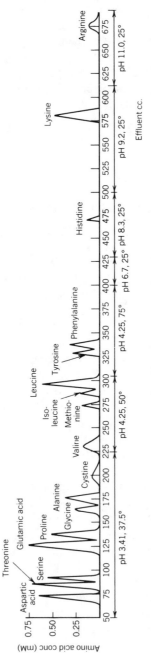

Figure 9.8. Classic ion exchange separation of amino acids using step-gradients on a 100-cm Dowex 50 column. Reprinted with permission from ref. 20. Copyright 1951, American Society of Biological Chemists, Inc.

called *amino acid analyzers*, and they were probably the first LC instruments, although they were not generally recognized as such.

Today there is a renewed interest in IEC, which has been caused by the developments in other LC techniques. However, the old resins and classical procedures are still used, so this section will be used to present the fundamentals of IEC and old methodology. The next section will describe the major advances resulting in high performance IEC.

In the preceding discussion, it was established that the following items were important in IEC:

1. The selectivity of the resin for various ions.
2. The particular ionic form (counterion) of the resin at the start of the analysis.
3. The concentration of this counterion in the mobile phase.
4. The pH of the mobile phase.

These items will be considered for the classical ion exchange resins.

Resins. The stationary phase or resin commonly used in IEC is the styrene/divinylbenzene copolymer shown in Figure 9.1. Divinylbenzene provides the crosslinking, and its percentage is usually specified in the resin specifications. Functional groups are put on the phenyl rings to provide the ionic sites; the most common ones are listed in Table 9, which includes both cation and anion types. The labels "strongly" and "weakly" refer to the acid/base strengths of the functional groups (acids for the cation

TABLE 9   Common Types of Ion-Exchange Resins

	Cation		Anion	
	Strongly Acidic	Weakly Acidic	Strongly Basic	Weakly Basic
Functional group	$-SO_3^- H^+$	$-CO_2^- H^+$	$\begin{array}{c} CH_3 \\ \mid \\ -N^+-CH_3\ Cl^- \\ \mid \\ CH_3 \end{array}$	$\begin{array}{c} R \\ \mid \\ -N^+-H\ Cl^- \\ \mid \\ R \end{array}$
Trade name				
Dowex	50W	—	1	3
Duolite	C-20	CC-3	A-101	A-2
Amerlite	IR-120	IRC-50	IRA-400	IR-45
Permutit	Q-100	Q-210	S-100	S-300

resins and bases for the anion resins). The weak exchangers can only be used over limited pH ranges: ≥ 6 for the cation resins and ≤ 8 for the anion resins. The number of ionic sites put on the resin will determine the capacity of the resin; typical values are 2–5 milliequivalents per gram of resin. Sample sizes must be chosen to be small enough not to exceed these limits.

The degree of crosslinking is usually between 4 and 16%. Highly crosslinked resins are harder, more brittle, less permeable, less susceptible to swelling and changing of bed volume, and more selective. A compromise around 8% is usually required to get sufficient permeability without excessive variations in bed volume with changing conditions.

Many other substrates have been used for IEC in addition to the polystyrenes. Some are natural products like cellulose, often crosslinked for added rigidity, and others are manufactured like polydextran gels. Most of them can only be used at low pressures, like the polystyrenes just discussed.

Operating Conditions. Table 10 shows typical selectivities of a cation resin as a function of crosslinking. For the monovalent cations given, Li is the least strongly sorbed, and the other values are calculated relative to it. Divalent cations are more strongly held then monovalent ions; for a typical resin, the order among cations is

$$UO_2 < Mg < Zn < Co < Cu < Cd < Ni < Ca < Sr < Pb < Ba$$

TABLE 10   Relative Partition Coefficients for Some Cations[a]

Cation	Crosslinking		
	4%	8%	10%
Li	1.00	1.00	1.00
H	1.30	1.26	1.45
Na	1.49	1.88	2.23
NH$_4$	1.75	2.22	3.07
K	2.09	2.63	4.15
Rb	2.22	2.89	4.19
Cs	2.37	2.91	4.15
Ag	4.00	7.36	19.4
Tl	5.20	9.66	22.2

[a] *Source:* Mallinckrodt Chemical Co.

For anions on an anion exchange resin, a typical order is

$$F^- < OH^- < \text{acetate} < \text{formate} < Cl^- < SCN^- < Br^-$$
$$< \text{chromate} < \text{nitrate} < I^- < \text{oxalate} < \text{sulfate} < \text{citrate}$$

These resins are not selective enough to permit the separation of ions that are close to one another in the listing. For example, Li, Na, and K cannot be separated.

However, a common cation separation that has become popular[21] uses the formation of coordination complexes to convert cations to anions via their chloro complexes:

$$Fe^{3+} \overset{Cl^-}{\rightleftharpoons} FeCl^{2+} \overset{Cl^-}{\rightleftharpoons} FeCl_2^+ \overset{Cl^-}{\rightleftharpoons} FeCl_3 \overset{Cl^-}{\rightleftharpoons}$$
$$FeCl_4^- \overset{Cl^-}{\rightleftharpoons} FeCl_5^{2-} \overset{Cl^-}{\rightleftharpoons} FeCl_6^{3-} \quad (2)$$

$FeCl_6^{3-}$ is formed in 12 M HCl, and the resulting anion is strongly held on an anion exchange column. Several divalent cations that form these chloro complexes to lesser degrees can be separated from iron and from each other: Ni, Mn, Co, Cu, and Zn. The sample is made up in 12 M HCl to convert the metals to the chloro complexes. Ni(II) does not form an anionic complex and can be washed off the column while the other metals are retained. Using a stepwise gradient elution with decreasing concentrations of HCl, the other chloro complexes are destroyed, one by one, reverting to cations that are not retained and are washed off the column. This separation depends on the equilibria involved in the chloro complex formations and not on chromatographic equilibria. Perhaps it should not be classed as a chromatographic separation for that reason and for the fact that it does not meet the usual chromatographic requirement that the species put on the column (the sample) is the same one that is eluted. In this case, anions are put on the column but cations are eluted.

But let us return to conventional IEC and consider the order of anion selectivities. It can be seen that an anion resin in the citrate form would be useless—citrate is the anion most strongly held, and it would not be appreciably displaced by any other anion in the list. To convert a citrate or sulfate resin to another form requires extensive washing with a mobile phase highly concentrated in the other ion in order to effect the displacement equilibrium. It is even difficult to convert completely a chloride resin to its hydroxide form. Thus, in purchasing a resin, it is important to note the counterion with which it is sold. Commonly anion resins are in the chloride form and cation resins in the hydrogen form.

Similarly, the counterion in the mobile phase during run can affect the partition ratios. Figure 9.9 shows an example of a separation of organic acids at pH 4.4 in which the only variable was the cation counterion— K^+ in Figure 9.9a and Na^+ in Figure 9.9b. Because K^+ is more strongly held than Na^+, and it competes with the analytes for ionic sites on the resin, the overall retention times in Figure 9.9a are shorter. A secondary effect in this case is the alteration in the elution order.

As with other LC processes, the slow diffusion in the mobile liquid phase is a source of zone broadening. Diffusion inside an IE resin in the presence of fairly high ionic strengths is especially slow. Traditionally IEC must be run very slowly to allow time for exchange to occur. The rates are increased at increased temperature, but classic separations are usually run at ambient temperature. Where an instrument is used and temperature can be controlled, the range of 40 to 60°C is usually beneficial.

When organic acids or bases are separated by IEC, the pH can be used to affect their degree of ionization and hence their partition ratios. This

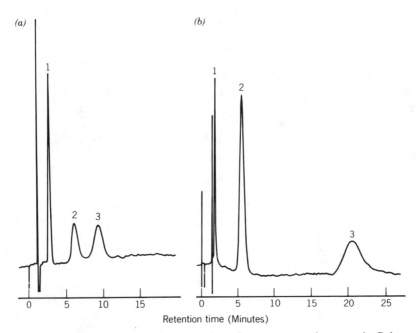

Figure 9.9. Effect of ion exchange counterion on separation speed. Column: Zipax SCX. (a) 0.1 M KH_2PO_4, pH 4.4; (b) 0.1 M NaH_2PO_4, pH 4.4. Reprinted from R. C. Williams, D. R. Baker, and J. A. Schmit, *J. Chromatogr. Sci.* **1973,** *11*, 619 by permission of Preston Publications, Inc.

is most effective within ±1 to 2 pH units of their pK values. Figure 9.10 shows a typical example of the effect of pH on retention time.

Gradient elution is also helpful; for acids, the pH can be increased during a run, hastening the elution of the stronger acids. Alternatively or in addition, the ionic strength can be increased to promote faster elution. Such a gradient system was used in the analysis of amino acids mentioned earlier (Figure 9.8).

Modern IEC. Improved stationary phases[22] similar to those developed for the other types of LC have led to improved separations by IEC. The old resins described above were followed by pellicular resins that were much more efficient and incompressible but had lower capacities. As is the case in the other forms of LC, they have been largely replaced by small microporous particles, silica and polymeric, that have the ionic groups directly on the particle or attached to a ligate or polymer on the particle surface.

These developments were spurred by new resins, produced by Dow Chemical in 1975 and licensed to a new company that was founded to exploit them for inorganic analysis.[23] In doing so, the company incorporated a second column in their instrument to remove ions that caused the mobile phase to have a high conductivity, thus permitting detection by conductivity; they used the term *ion chromatography* (*IC*) to distinguish it from other IE instruments. Although the name *ion chromatography* has continued to be associated with this form of IEC, and the

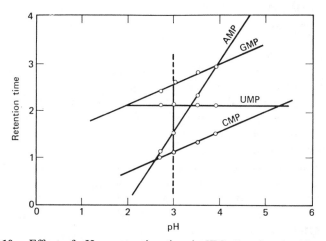

Figure 9.10. Effect of pH on retention time in IEC. Reprinted with permission of Alltech Associates/Applied Science Labs, Deerfield, IL.

manufacturers who make this equipment contend that it is a special technique, the separations performed by IC are ion exchange separations.

Ion Chromatography (IC). Aside from the new ion exchangers developed for IC, the major difference between it and other IE techniques is the use of a second column to suppress the ionic strength in the mobile phase so that the analyte ions can be detected conductimetrically.[24] Originally this was achieved by using, as part of the mobile phase, a counterion that could be converted to a molecular form by an ion exchange process in the second, suppressor, column.[25] For example, in an anion analysis using carbonate as the counterion, nonionic carbonic acid is formed when the original cations (K^+ or Na^+) are exchanged for H^+ in the suppressor column, which is a strong cation exchanger. The acidity of the mobile phase helps prevent the ionization of the carbonic acid without affecting the analytes that are conjugate bases of strong acids—ions like nitrate, chloride, and bromide. Figure 9.11 shows a diagrammatic representation of the separation of nitrate and sulfate ions.

The use of a suppressor column is not without problems. Eventually the resin becomes exhausted and needs to be regenerated, which is inconvenient. Also, the slightly ionized carbonic acid produces a small continuous baseline conductance signal, so that the vacancy peak from the sample injection is detected and produces a negative peak that can interfere with other analytes eluting at that time. The use of a second column also results in some zone broadening, of course, which decreases the overall efficiency of the analysis.

The first problem has been resolved using continuous, flowing suppressor streams that contact the chromatographic stream through a porous membrane. Hollow-fiber ion exchange tubing (Du Pont's Nafion™) packed with plastic beads to decrease the internal volume and zone spreading has been used.[26] Figure 9.12 shows the action of two designs: (*a*) a hollow fiber, and (*b*) a sandwich.

The vacancy peak (carbonate dip) has been eliminated as a problem by using faster, more efficient IE columns[27] so that the vacancy peak occurs early in the chromatogram.

The desire to use a universal detector like a conductance detector is the reason for the use of a second, suppressor, column. Alternative detector systems have been used without a suppressor; some ions absorb in the UV and can be used with non-UV absorbing counterions in the mobile phase; some work has been done with the universal refractive index detector; and, finally, vacancy chromatography (discussed more fully later in this chapter) has been applied whereby the mobile phase

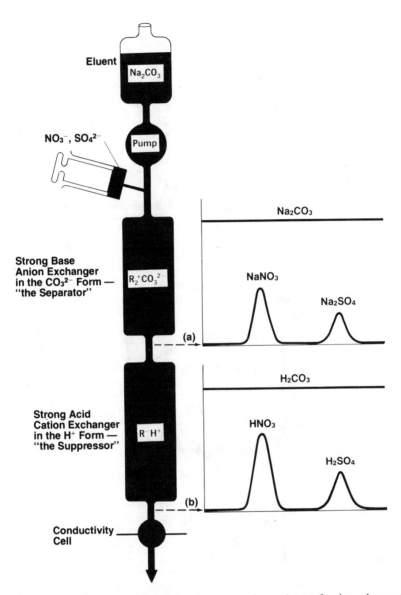

Figure 9.11. Representation of eluent suppression scheme for ion chromatographic separation of nitrate and sulfate. Reprinted with permission from H. Small, *Anal. Chem.* **1983**, *55*, 235A. Copyright 1983, American Chemical Society.

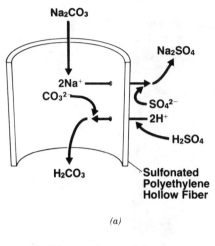

$$Na_2CO_3$$

$$2Na^+ \quad \bullet \multimap$$
$$CO_3{}^{2-}$$

$$Na_2SO_4$$

$$SO_4{}^{2-}$$
$$2H^+$$

$$H_2SO_4$$

$$H_2CO_3$$

Sulfonated
Polyethylene
Hollow Fiber

(a)

**Eluant Flowpath in
MicroMembrane Suppressors**

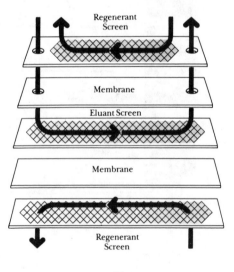

Regenerant
Screen

Membrane

Eluant Screen

Membrane

Regenerant
Screen

(b)

Figure 9.12. Alternative forms of suppressor column for ion chromatography: (*a*) hollow fiber, (*b*) membrane sandwich. Reprinted by permission of Dionex Corporation, Sunnyvale, CA.

contains a UV absorber like phthalate ion and the analyte peaks are detected as negative peaks by a UV detector.[28]

Most important, the improved technology has produced excellent separations of ions that could not be directly separated by IEC, as was noted earlier. Some inorganic separations are shown in Figures 9.13 to 9.15. Organic ions can also be separated, of course, but there is another important special type of analysis of ions that is popular for organic analysis. It is called *ion-pair* chromatography.

Ion Pair Chromatography (IPC). This discussion is limited to the most common type of IPC—reverse phase mode using bonded alkyl ligates (like C_{18} and C_8) as the stationary phase. Its main advantage over regular reverse phase LC is that it facilitates the analysis of samples that contain both ions and molecular species.

In regular reverse phase operation, ionic species are not retained much if at all, and pH is used to control the partition ratios by controlling the degree of ionization. If a sample contains analytes that vary widely in pK values, only a few of them can be separated at a given pH, and the others (as ions) will elute unseparated close to the dead volume (nonretained) peak. Several groups of workers suggested the addition of counterions to the mobile phase to form neutral ion pairs with the analyte ions, thus causing them to be attracted to the stationary phase, retained, and separated. Schill and his group were working in ion-pair extractions and adapted their procedures to chromatography,[29] which became known as

Figure 9.13. Separation of common anions by ion chromatography. Reprinted by permission of Dionex Corporation, Sunnyvale, CA.

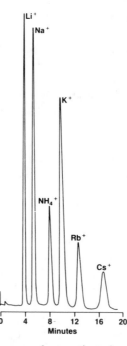

Figure 9.14. Separation of monovalent cations by ion chromatography. Reprinted by permission of Dionex Corporation, Sunnyvale, CA.

extraction chromatography. Knox[30] and his group used the detergent cetyltrimethylammonium bromide in the mobile phase and dubbed their technique *soap chromatography*. Other workers entered the field, which became confused by the diversity of names by which it was called and by the debate that arose over the mechanism of action.

In summary, two mechanisms have been proposed to explain the results. One assumes that the ion pairs are formed in the mobile phase and behave as nonionic moities similar to other polar molecules in RPLC. The other rests on the belief that the counterions selectively sorb in the stationary phase and attract and retard the analyte ions by an ion exchange mechanism; this mechanism requires that the ion-pair reagent have a hydrophobic end that would be attracted to the alkyl chain on the bonded phase and an ionic site on the other end. Perhaps both mechanisms are partially correct, and the predominant mechanism may depend on the operating conditions. In any case the following discussion will not attempt to distinguish between the mechanisms or justify either one.

The counterions commonly used in IPC are listed in Table 11. Waters Associates introduced a line of reagents called PIC™ reagents, and these

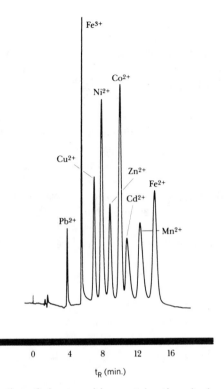

Figure 9.15. Separation of nine transition metal cations by ion chromatography. Reprinted by permission of Dionex Corporation, Sunnyvale, CA.

TABLE 11   Some Ion-Pair Reagents/Counterions

For Basic Samples	For Acidic Samples
Methanesulfonic acid	Tetrabutylammonium hydroxide
Pentanesulfonate: Na salt	Tetraethylammonium hydroxide
(PIC reagent B-5)	Tetrabutylammonium phosphate
Hexanesulfonate: Na salt	(PIC reagent A)
(PIC reagent B-6)	Hexadecyltrimethylammonium
Heptanesulfonate: Na salt	bromide
(PIC reagent B-7)	Trihexylamine
Octanesulfonate: Na salt	Triheptylamine
(PIC reagent B-8)	Trioctylamine
2-Naphthalenesulfonate: Na salt	
Dodecylsulfate: Na salt	
Dioctylsulfosuccinate: Na salt	
Citric acid	
Picric acid	
Perchloric acid	

names are included in the table. In using the ion pair reagents, the charge
on the reagent must be opposite to the charge on the analytes, obviously.
The amines that are listed are not ionic, but they are used in acidic situations where the protonated (cationic) forms exist. Increasing the concentration of ion pair reagent will increase analyte partition ratios up to
a point and level out. Typical concentrations are 0.005 to 0.05 M, although
higher concentrations are sometimes required.

As IPC is currently practiced, a bonded reverse phase separation is
usually the first system tried. If some ions cannot be retained, an ion pair
reagent is added. A typical separation is shown in Figure 9.16.

Alternatively, the use of an ion pairing reagent can improve a chromatogram by decreasing the retention times and improving the peak
shapes, as shown by Bidlingmeyer[31] (Figure 9.17).

In general, IPC is a useful technique as an alternative to IEC for the
separation of ions or mixtures of ions and molecules. It does have some

PACKING: μBONDAPAK/C$_{18}$
                PIC^{TM} SEPARATION
COLUMN: 4 mm x 30 cm
SOLVENT: Methanol/Water (50/50) PIC Reagent B-7

1 Maleic Acid
2 Phenylephrine
3 Phenylpropanolamine
4 Phenacetin
5 Naphazoline
6 Pyrilamine

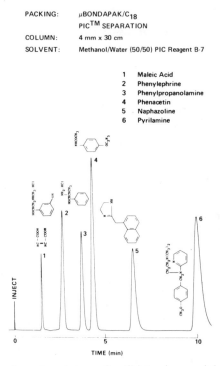

Figure 9.16. Separation of a mixture of antihistamines and decongestants by ion
pair chromatography. Courtesy of Millipore Corp., Waters Chromatography
Division.

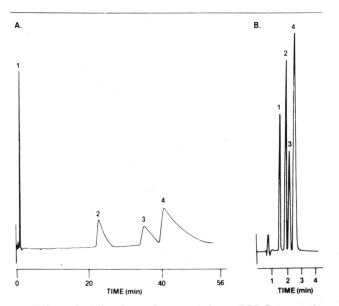

Figure 9.17. Effect of adding ion pair reagent in an RPLC separation of anesthetics. Samples: (1) benzocaine (peak 2 in chromatogram *b*; (2) lidocaine (peak 1 in chromatogram *b*); (3) tetracaine; (4) etidocaine. Column: Radial Pak C_{18}. Mobile phase: 60% acetonitrile in water; pH 3; 1% triethylamine added in chromatogram *b*. Reprinted from the *Journal of Chromatographic Science* by permission of Preston Publications, Inc.

disadvantages: the ionic solutions are often corrosive and result in short column life, some of them also absorb in the UV and limit the use of the UV detector, and the silica-based supports are limited to pH values below about 7.5. In addition, the mobile phases should not be left standing overnight. They should either be replaced by water or pumped continuously at a slow rate. Further details about IPC can be found in the review and book by Hearn.[32] Suggestions for column care have been given by Rabel.[22,33]

Ligand Exchange Chromatography. A cation exchange resin can also be used to separate analytes that can form coordination complexes with the metal attached to the resin. For example, a cation resin in the Cu(II) or Zn(II) form can be used to separate amino acids using ammonia as a competing ligand. Some typical examples are given in references 34 and 35. Ligand exchange methods include separations achieved using ion pair conditions, and they can also be used for chiral separations (Chapter 11).

Summary. The discussion of modern IEC has included a number of peripheral techniques, but the basic ion exchange concept is the same as that described earlier. The major variables of pH, temperature, and counterion are used to advantage, as described earlier. In addition, an organic solvent is sometimes added to the mobile phase since it can have a significant effect on the separation. The effect is unpredictable, but not unexpected, especially for organic ions.

The fast, highly efficient resins used have smaller capacities than the old polystyrene resins. Values range from about 1 to 500 microequivalents per gram of resin, which is about one-tenth to one-thousandth of the old resins.

From the examples presented it can be seen that IEC has developed into a widely versatile and useful complement to the other forms of LC. For example, the separation of amino acids in Figure 9.18 is a vast improvement over the original work just 30 years earlier (Figure 9.8), and ion separations like those in Figures 9.13 to 9.15 were not possible until recently.

Size Exclusion Chromatography (SEC)

This technique is different from all the other LC methods because separations are based on a physical sieving process, not on chemical attrac-

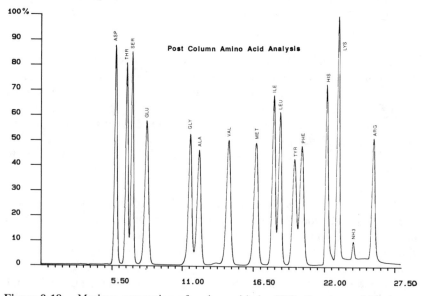

Figure 9.18. Modern separation of amino acids by IEC. Courtesy of Kratos.

tions and interactions. Therefore, the purpose of the mobile phase is merely to act as a solvent for the sample.

SEC has its origin in two separate groups of workers and was called by two separate names. Beginning in 1959 the name *gel filtration chromatography* came into use by those using dextran gels to separate biochemical polymers using aqueous mobile phases, while, a few years later, the name *gel permeation chromatography* was coined by polymer chemists separating synthetic organic polymers on polystyrene gels using nonaqueous mobile phases. Other names were also used, but now it seems preferable to use the term *size exclusion* to include both types, and the older terms should be discarded.

SEC is a method for separating molecules based on their sizes by using stationary phases with pore sizes capable of discriminating among the analytes in a sample. Its primary application has been to the analysis of polymers, and it is commonly used to get molecular weight distributions of polymers.

Theory. To a first approximation, the theory is very simple. A sample containing analytes with a variety of sizes is introduced to the SEC column, which is packed with a stationary phase that is inert except that it has pores of about the same size as the analytes. The largest molecules will be unable to penetrate the pores, will be excluded, and will elute at the time for nonretained materials. Slightly smaller molecules will be able to penetrate some of the pores and will be slightly retained. Smaller and smaller analytes will penetrate an increasing fraction of the pores in the packed bed and be increasingly retained and thus separated from each other. Eventually an analyte size is reached that can penetrate all the pores, and all molecules of that size and smaller will elute together and represent the limit of the column's separating power. This behavior is represented in Figure 9.19, which shows the total exclusion and total permeation limits. Note that the x axis is logarithmic and that the large molecules are eluted first, followed by the smaller ones.

The basic chromatographic equation

$$V_R = V_M + KV_S \tag{3}$$

takes on new meaning for SEC. The volume of the stationary phase, V_S, is really the volume of the pores of the stationary phase, and the partition coefficient K has a new meaning. It measures the extent to which a given analyte has penetrated the pores and has a value of zero for total exclusion and unity for total permeation. This restriction on the possible values of K severely limits the range of analytes that can be separated on one SEC column and verifies the prediction in Chapter 1 that SEC has a maximum peak capacity of about 20 analytes base to base. This corresponds to a range in molecular weights of about 1.5 decades, as shown in Figure 9.19. In actuality, some adsorption of analytes on the stationary phase often occurs, and partition coefficients greater than 1 are observed.

Also, the theory is much more complex than just presented. Bly[36] has classified the process into three mechanisms: steric exclusion, restricted diffusion, and thermodynamic considerations, and the process has been thoroughly studied. The rate equation is also different for SEC. For polymers, the longer retained peaks have smaller peak dispersivities, H, than early peaks, in direct contrast to normal LC expectations. In part this is due to the fact that the smaller molecules that elute last have higher diffusion coefficients and therefore less mass transfer zone spreading.

Stationary Phases. The stationary phases used in SEC are solids ranging from synthetic polymers to glasses. The earliest ones were soft gels, which cannot be used at high pressures. Consequently, the development of new phases has paralleled somewhat the development of more rigid resins and gels in IEC, as noted earlier. Some phases and their trade names are given in Table 12.

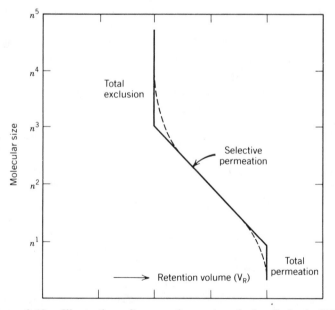

Figure 9.19. Illustration of permeation and exclusion limits in SEC.

The rigid phases, preferred for fast analysis, are chosen based on their pore sizes (and range of applicability), their compatibility with the mobile phase to be used, and their inertness. Since most of them are made from silica or glass, they are restricted to pH values less than about 7.5. If adsorption is a problem, the phases can be deactivated; the silylation reactions discusssed earlier are popular.

Rigid packings with smaller pores have been developed more recently to permit the separation of smaller molecules with molecular weights in the range of 100 to 1000. They are fast and capable of baseline resolution of small homologs.

Since any single column packing has a limited range, it is often necessary to perform SEC with multiple columns in series. This procedure extends the molecular weight range over which separations can be effected.

Mobile Phases. There are few requirements the mobile liquid must meet: it should (1) be a good solvent for the sample; (2) have a low viscosity, like any LC solvent; (3) wet the stationary phase and, if possible, help deactivate it; and (4) be compatible with the detector being used. For soft gels, it must also cause the gel to swell. In general, polar liquids like water, THF, and chloroform are used.

Applications. SEC is the easiest LC technique to understand and apply. Choosing the stationary and mobile phases is relatively easy as just described, and the separation achieved on the selected system is predictable.

It is used mainly to separate large molecules, synthetic polymers, and biopolymers with molecular weights between 2,000 and 2,000,000. Figure 9.20 shows a typical separation of a surfactant in which the individual

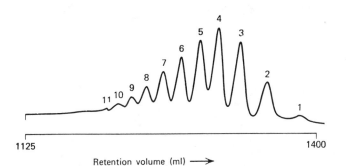

Figure 9.20. SEC separation of oligomers in a commercial surfactant (Triton X-45). From K. M. Bombaugh, W. A. Dark, and R. F. Levangie, *Separ. Sci.* **1968,** *3,* 375; courtesy of Marcel Dekker.

TABLE 12 Commercially Available Packings for SEC[a]

Principal Tradename[b] (Supplier)	Chemical Type/ Particle Size	Flow Limits: Linear Flow Rate[c] Pressure Drop[d]	Typical Efficiency (plates/ft)[e]	Upper MW Limit (Calibrator)	Typical Assay Time[f]
		Polymeric, rigid			
TSK PW (Toyo Soda) *Spherogel PW (Altex) *Bio-Gel Tsk (Bio-Rad)	Cross linked hydroxylated polyether 11–15 μm	160 cm/hr 200 psi	3500	3×10^7 (Dextran)[g]	30 min
Shodex Ionpak (Showa Denko) *Shodex Aqueous (Perkin Elmer)	Sulfonated crosslinked polystyrene 10 μm	170 cm/hr 500 psi	5000	5×10^6 (Dextran)[g]	15 min
**Spheron (LaChema) (Koch-Light) (Knauer KG)	Crosslinked poly (2-hydroxyethyl-methacrylate) 20–40 μm	100 cm/hr 500 psi	800	1×10^7 (Dextran)[g]	60 min
		Polymeric, compressible			
*Toyopearl (Toyo Soda) *Fractogel (E. Merck)	Crosslinked hydroxylated polyether 30–60 μm	25 cm/hr	800	1×10^7 (Dextran)[g]	150 min
**Sephadex (Pharmacia)	Crosslinked dextran 40–120 μm (dry)	10 cm/hr	200	2×10^5 (Dextran)	—
**Sephacryl (Pharmacia)	Sephadex, sequentially reacted with allyl chloride and methylene bisacrylamide	20 cm/hr 14 psi	500	1×10^8 (4000 Å) (Dextran)[g]	90 min

**Ultrogel AcA (LKB)	Agarose, postpolymerized with methylene bisacrylamide 25–55 μm (dry)	25 cm/hr 2 psi	500	2×10^6 (Protein)	100 min
**Bio-Gel A (Bio-Rad)	Agarose 40–80 μm; 80–150 μm; 150–300 μm (dry)	10 cm/hr 2 psi	200	1×10^8 ?	250 min

Siliceous, underivatized

**Li Chrospher (E. Merck) Du Pont SEC	Porous silica 10 μm	1800 cm/hr 1000 psi	7000	5×10^4 (Dextran)[g]	10 min
Zorbax SE (9–11 μm) Zorbax PSM (5–7 μm)	Porous silica 5–11 μm	4000 cm/hr —	2500	5×10^6 (Dextran)[g]	10 min
**Spherosil (Rhone-Progil)	Porous silica 40–110 μm	— —	300	—	100 min
**CPG (Electronucleonics)	Porous glass 55, 100, 150 μm (all ± 30%)	300 cm/hr —	250 700[h]	3000 Å	60 min

Siliceous, derivatized

**LiChrosorb Diol (E. Merck)	1,2-dihydroxypropyl silica 5 μm	400 cm/hr 2000 psi	5000	2×10^6 (Dextran)[g]	10 min
SynChropak (Synchrom)	1,2-dihydroxypropyl silica 10 μm	400 cm/hr 2000 psi	3500	1×10^7 (Dextran)[g]	10 min
Aquapore (Brownlee)	1,2-dihydroxypropyl silica 10 μm	400 cm/hr 2000 psi	3500	1×10^7 (Dextran)[g]	10 min
Aquachrom (Chromatix)	1,2-dihydroxypropyl silica 10 μm	400 cm/hr 2000 psi	3500	1×10^7 (Dextran)[g]	10 min

TABLE 12 Commercially Available Packings for SEC[a] (*continued*)

Principal Tradename[b] (Supplier)	Chemical Type/ Particle Size	Flow Limits: Linear Flow Rate[c] Pressure Drop[d]	Typical Efficiency (plates/ft)[e]	Upper MW Limit (Calibrator)	Typical Assay Time[f]
TSK SW (Toyo Soda)	Derivatized silica treatment unknown 8–12 gm	300 cm/hr 800 psi	5000	6×10^5 (Dextran)[g]	30 min
μBondagel (Waters)	Derivatized silica treatment unknown 8–12 μm ("Polyether phase")	500 cm/hr 1200 psi	3500	1×10^7 (Dextran)[g]	—
**GlycophaseG (Pierce) *Glycerol GPC (Electronucleonics)	1,2-dihydroxypropyl glass 55, 100 (±20) μm	300 cm/hr —	200	—	50 min

[a] Reprinted from P. L. Dubin, *Amer. Lab.*, *1983*, **15**(1), 62. Copyright 1983 by International Scientific Communications.
[b] * Commercial name. ** Substrate sold as bulk packing; optimal data from literature.
[c] Limitation on linear flow rate as imposed by loss of efficiency, with or without gel compression.
[d] Pressure drop corresponding to flow rate in (c), or maximum encountered in normal operation (30 cm column).
[e] Obtained with low MW nonionic solute
[f] Values found in published literature.
[g] Extrapolated from data on lower MW dextrans, or inferred from results with larger particles, e.g, viruses.
[h] Column packed with sieved glass.

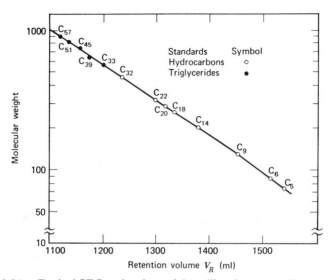

Figure 9.21. Typical SEC molecular weight calibration curve. From K. M. Bombaugh, W. A. Dark, and R. F. Levangie, *Separ. Sci.* **1968,** *3,* 375; courtesy of Marcel Dekker.

oligomers have been resolved. Cazes[37] listed 100 *types* of compounds that have been studied by SEC.

To determine molecular weights, a calibration curve like that in Figure 9.21 is first prepared. This is a semilog plot like the general one shown earlier (Figure 9.19) and is linear from C_9 to C_{57}, including both hydrocarbons and triglycerides.

When SEC curve is used to determine the molecular weight distribution of a polymer, one single large peak is usually obtained. Higher resolution is neither necessary nor desirable because the desired information can be obtained from the peak shape (cumulative area vs. retention), as described by Cazes.[38]

SEC has become an important part of polymer analysis, and further details can be found in books on that topic as well as books on SEC.[39]

INSTRUMENTATION

A schematic diagram of the apparatus used for high performance LC is shown in Figure 9.22. Not all the parts shown are required for a simple apparatus, but most are included in commercial instruments. More information about commercial instruments can be found in the review by McNair.[40]

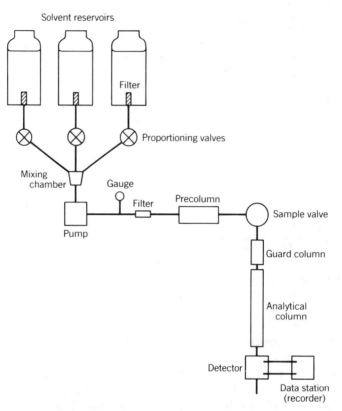

Figure 9.22. Typical LC apparatus.

Pumping Systems and the Mobile Phase

Before discussing the actual pumps, let us consider the mobile phase and its requirements. The mobile liquid must be very pure, and special chromatographic grades are available for most common solvents, including water. Some solvents are routinely stabilized with small amounts of chemicals that can significantly alter their solvent properties for LC use and may absorb in the UV. For example, chloroform is often stabilized with ethanol or pentene. Even particulate matter finds its way into pure solvents requiring the use of a filter in the intake line from the solvent reservoir.

Because LC liquids are chosen for their low viscosity and high volatility, they are often fire hazards if they are at all flammable. Others are toxic. Care must be exercised in their use; consult standard safety manuals. Like most ethers, THF and dioxane tend to form explosive peroxides. They should not be stored for long periods of time, and care must be exercised if they are purified. Before attempting any distillation, consult a reliable reference such as the one by Riddick and Bunger.[41]

Air is soluble in many solvents, including water, and it should be removed prior to use because it tends to be released at the low pressure end of the apparatus, the detector, where it causes noise or even a complete loss of signal. Many commercial instruments include the degassing capability in the solvent reservoir for convenience. Some possibilities are the following, used individually or in combination: vacuum, heating, sonication, and sparging with helium. The latter is probably the easiest to include in an instrument, but the most efficient is the combination of sonication under vacuum.

Pumps. The high pressure pump required for LC is a critical part of the apparatus, so a lot of attention has been directed toward the development of good pumps. A versatile pump should be able to deliver flows from a fraction of a milliliter up to at least 10 mL/min with an accuracy of about 1% up to a maximum pressure around 35 MPa. Two types can be distinguished: constant flow and constant pressure. Constant flow is most desirable for the concentration type detectors and for reproducible retention times, but these pumps often produce noise in the detector because of the cyclic nature of their action. Remember that a blockage in the LC system will cause a constant flow pump to produce increasingly high pressures, necessitating protection with a pressure relief valve. It will cause a constant pressure pump to deliver decreasing flows.

The reciprocating action pump shown in Figure 9.23 is the most common type in use, and it delivers a constant flow. The pulsations that result can be removed or lessened by damping, but a better solution has been the use of two heads out of phase with one another, as shown in Figure 9.24. Sophisticated, computer-designed camshafts have been built to minimize the pulsations from these pumps, and additional dampening is usually not required.

The other type of constant flow pump is a positive displacement syringe. It is pulseless but suffers from the limited volume it can deliver before refilling. The major type of constant pressure pump is a pneumatically driven syringe.

Gradient Devices. Gradient devices can also be divided into two types: high pressure and low pressure. In a typical low pressure system (depicted in Figure 9.22), various proportions of up to three solvents are mixed before they enter the high pressure pump. The proportions are regulated by valves that are in turn controlled electronically with microprocessors.

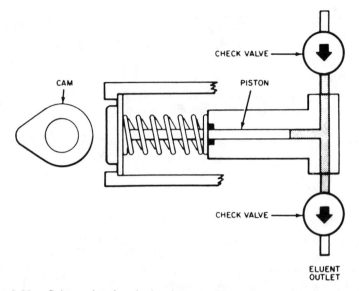

Figure 9.23. Schematic of a single piston reciprocating pump. Reprinted from Yost, Ettre, and Conlon, *Practical Liquid Chromatography*, 1980 by courtesy of Perkin-Elmer.

Some means must be provided to mix the liquids thoroughly before they enter the pump, and a typical mixing chamber of low volume is shown in the figure.

High pressure systems provide one high pressure pump for each solvent; the output from each pump is regulated by an electronic control, and the combined streams are mixed and sent to the sample valve. Although this system is more costly because of the need for two pumps, it produces more precise mixtures at the extremes of a gradient.

In all gradient devices, the tubing volumes must be kept small so the gradient formed is quickly delivered to the column. Prior degassing is required because the mixing process often results in the liberation of heat, which can result in gas evolution since the solubility of dissolved gases is lower at the higher temperature. The heating caused by the entropy of mixing in itself can be a problem unless temperature control is provided. Another concern in designing gradient systems is the change in viscosity that accompanies mixed solvents. Figure 9.25 shows the nonlinear behavior of mixtures of some common solvents with water. As a minimum, constant flow pumping systems are required to handle these gradients.

The gradient system should be capable of producing a variety of gradient shapes. Linear gradients are common, but the Snyder solvent parameter produced will not likely be linear, as was shown earlier in Figure

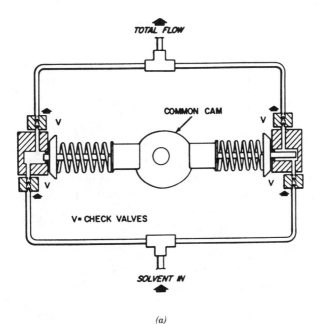

(a)

(b)

Figure 9.24. (*a*) Schematic and (*b*) flow profiles of a dual-head reciprocating pump. Reprinted from Yost, Ettre, and Conlon, *Practical Liquid Chromatography*, 1980 by courtesy of Perkin-Elmer.

9.2. To get a linear *polarity gradient* with the systems shown in Figure 9.2 would require a concave gradient. Choosing a gradient is often done by trial and error using an initial rate of 2% per minute. A variety of references are available for further details.[42-44]

Sampling

Figure 9.21 included a precolumn and a filter in the line before the sample valve. These are not required, but the filter is highly recommended. The precolumn may be necessary to precondition the solvent before the sam-

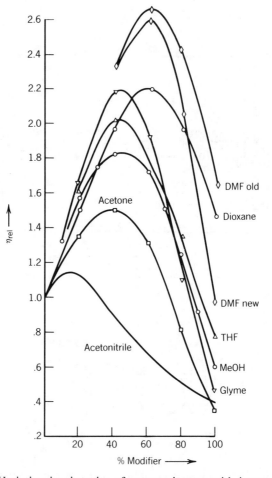

Figure 9.25. Variation in viscosity of water mixtures with increasing amounts of modifiers. Reprinted with permission from S. van der Wal, *Chromatographia* **1985,** *20,* 274.

ple is introduced and is mandatory for LLC. When the stationary phase is silica or silica based, a precolumn of silica is often used to saturate the mobile phase with silica and thus decrease the dissolution of silica at the top of the column. This is especially recommended when working near the upper pH limit of silica.

High pressure valves are most often used for sampling in LC, although syringe injection similar to that used in GC was formerly employed. A typical valve has six ports and two positions, one for loading and one for sample introduction to the column. Interchangeable loops with different calibrated volumes are attached to the valve and are filled with a syringe. They range in volume from 10 μL to several hundred microliters. For smaller volumes, special valves have been designed with internal loops and low volume connections. Alternatively, the sample loop can be only partially filled with sample. Many details need to be considered for good sampling, and the paper by Dolan[45] is highly recommended.

Other possibilities include *stop-flow* injection and septumless syringe injection valves, in which the injection into the valve is made off-line under ambient pressure.

The sample should be dissolved in the mobile phase. If it is not very soluble, other solvents can be used, but they should be as similar to the mobile phase as possible, and blanks should be run to see the effect of this new solvent on the column. A better alternative is to use larger sample volumes. The upper limit is about one-thirtieth of V_M, which would be about 100 μL for a conventional analytical column.

Columns

The column packings were described in the respective sections already discussed. This section deals with column sizes and other operational details.

Column Dimensions. The conventional columns we have been discussing have inside diameters of 4.6 mm, lengths of 10 to 25 cm, and are packed with porous microparticles 5 to 10 μm in diameter. Slightly wider or narrower columns are not much different in efficiency, and 4.6 mm is convenient because it is the i.d. of commercially available one-quarter-in. stainless steel. However, there is evidence that a *wall effect* decreases efficiency; when the injected analyte spreads out and reaches the wall of the column, it experiences differences in flow patterns with a subsequent spreading of the zone. It has been shown that, depending on the column length and the particle size, columns with diameters in the range of 5 to 8 mm are wide enough to serve as *infinite diameter* columns and have greater efficiencies since the sample elutes before it reaches the column wall.

Since efficiency can be increased by improving the density of the packing, several commercial units use special means for compressing the column after it is packed. Both radial and axial compression are available. Other columns have been designed for use with soft gels that tend to swell and shrink during use. They have movable end pieces that can be adjusted to conform to the bed volume. It is also becoming popular to pack columns in replaceable cartridges.

Fast LC. Some very short columns that are only 3 to 5 cm long and are packed with very small particles (3 μm) have become popular. Sometimes referred to as "3 × 3" columns, they are less costly and give good separations with minimum consumption of mobile phase. The term *fast or high-speed LC* has been applied to their use at high flow rates around 4 mL/min. Figure 9.26 compares a conventional column with a 3 × 3 column for the separation of six explosives. In general, separations that need 4,000 or fewer plates can be separated with these columns in a minimum of time with conventional detectors. The instrument requirements are more severe, and special detectors may be needed; some of the practical aspects of fast LC have been described by van der Wal.[46]

Microbore Columns. As the name implies, these column types have small inside diameters. The sizes vary, and some columns are packed while others are open tubes. A summary of the major types is presented in Table 13. The most common commercial columns are 1 mm i.d. and packed. The interest in these columns and the wide range of types has already resulted in the publication of three books on this topic alone.[47-49]

The advantages of microcolumns are: (1) high efficiencies, (2) increased detectivity, primarily with mass flow detectors, (3) decreased solvent consumption and lower costs, (4) compatibility with on-line detectors like MS, and (5) the possibility of using exotic mobile phases or reagents and special new detectors. The separation shown in Figure 9.27 is taken from a recent paper[50] and illustrates several of the desirable features just listed: the minimum plate height was only 16 μm; regular one-sixteenth-inch ss tubing was used (1.2 mm i.d.) for columns that were slurry packed; the flow rate was 0.2 mL/min (optimum was 0.1 mL/min), which required a small amount of solvent; analysis time was less than 6 min; detection limits were about 1 pg; and a specially constructed amperometric detector was used.

The disadvantage is the instrumental requirement to keep all "dead" volumes and detector volumes sufficiently small to prevent extracolumn zone broadening. It is also possible that special pumps, sample valves, and detectors may be required. The same requirements apply to the short columns used in fast LC.

In addition to the books mentioned above, several review articles provide additional details,[51,52] and the June 1985 issue of the *Journal of Chromatographic Science* is devoted to this topic.

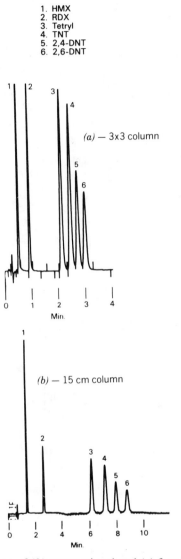

1. HMX
2. RDX
3. Tetryl
4. TNT
5. 2,4-DNT
6. 2,6-DNT

(a) — 3x3 column

(b) — 15 cm column

Figure 9.26. Comparison of (b) conventional and (a) fast 3 × 3 columns for the separation of explosives. Columns: Supelcosil LC-8. Reprinted from *Supelco Reporter*, Vol. IV, No. 1, February 1985 with permission of Supelco, Inc., Bellefonte PA.

TABLE 13 Column Characteristics

Type	i.d.	Length	Flow Rate	Packing Diameter
Conventional, packed	4.6 mm	3–25 cm	1 mL/min	5 μm
Microbore, packed	0.2–1 mm	10–10³ cm	1–20 μL/min	5 μm
OT, packed[a]	40–80 μm	1–100 m	0.5–2 μL/min	10–30 μm
OT, not packed	15–50 μm	1–100 m	<1 μL/min	

[a] Packing may be modified because columns are prepared by filling a glass capillary and then drawing it out, exposing the packing to high temperature and embedding it in the glass wall.

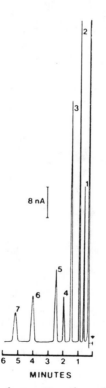

Figure 9.27. High speed microbore separation of biogenic amines. Reprinted with permission from E. J. Caliguri, P. Capella, L. Bottari, and I. N. Mefford, *Anal. Chem.* **1985,** *57,* 2423. Copyright 1985, American Chemical Society.

Other Column Features. *External Zone Broadening.* If attention is not paid to the extracolumn volumes in connecting tubes, injection devices, and detectors, the efficiency of a separation may be decreased by zone broadening in these regions. This is especially important for the small and microbore columns just described when they are used at low flow rates.

If we express the zone broadening in terms of the peak width, the equation that governs the total effect is

$$w_{tot}^2 = w_{col}^2 + w_{app}^2 \qquad (4)$$

where w represents the width of the peak at its base. The width of a peak due to the apparatus, w_{app}, can be determined by removing the column from the apparatus, joining together the injector and the detector, and running a one-component "sample." The width of the resulting peak can be measured, but fast detector and readout response times are required. A typical value could be 40 μL.

In order to estimate the effect of this extracolumn broadening on the overall (total) peak width, let us assume that a good column will produce peaks with widths as narrow as 100 μL. Then,

$$w_{tot}^2 = (100)^2 + (40)^2 = 11{,}600 \qquad (5)$$

and

$$w_{tot} = 108 \ \mu L \qquad (6)$$

Thus, the extra broadening of 40 μL has increased the peak width by 8 μL or 8%, which is not too bad.

Normal liquid chromatographs use small bore one-sixteenth-in. tubing (<0.25 mm i.d.), special connectors, minimum tube lengths, and detectors with volumes around 8 to 12 μL. For the low flow situations described earlier, the i.d. of the tubing is more critical and special detectors may be required; in the paper mentioned above,[50] the column was connected directly to the injection valve and the detector, thus eliminating all extracolumn tubing.

Guard Columns. To help protect analytical columns from degradation by dirty samples, it has become common to include a "guard" column between the injector and analytical column. These columns are short and intended to be discarded or repacked after they become contaminated. Often they are dry packed with a pellicular support that is chemically similar to the one being used in the analytical column. Although they contribute a little zone broadening to the system, the effect is small and worth the convenience of protecting the main column.

Temperature Control. In most of the early instruments, the columns were exposed to ambient conditions, and no provision was provided for thermostating them. More recently it has become clear that there are advantages to controlling the temperature, and some analyses are markedly improved at elevated temperatures. For example, higher temperatures provide faster kinetics, lower solvent viscosities, and decreased adsorption. Newer instruments provide this capability, and thermostatic jackets are commercially available for retrofitting on older ones.

Detectors

The most common LC detectors are based on ultraviolet absorption (UV), refractive index (RI) changes, or molecular fluorescence emission. All three are very common phenomena, and the detectors used in LC are, with few exceptions, modifications of existing technology and not based on new principles. This was not the case with GC detectors, many of which were invented specifically for GC. In fact, many of the GC detectors, such as the FID, electron capture, and TID, have been adapted to LC as well.

Of the three popular detectors, UV is the most popular—used in 71% of the papers in 1980–1981[53]; its use is limited mainly because it is not universal. The RI detector is universal, but it is not as sensitive or as stable. Thus the search continues for a sensitive, universal LC detector[53] to complement existing specific detectors.

The principles of UV absorption, RI, and fluorescence will not be discussed here; they are thoroughly treated in books on instrumental methods. Their classifications according to the definitions established in Chapter 7 are important and are included in Table 14. Note, for example, that all three are concentration-type detectors, which makes them susceptible to errors due to flow changes.

The major modification required to adapt conventional detectors for use in LC is to design flow cells that have small volumes. We have already seen that volumes around 10 μL are required in most cases to prevent peak broadening. The geometry of the flow channel is also critical in achieving a maximum path length without undue sensitivity to flow oscillations, turbulent flow, or (in the case of UV) refractive index changes.

UV Detectors. The original UV detectors for LC used a low pressure mercury lamp as a source and were designed to be used at a single fixed

TABLE 14 Properties of Common LC Detectors

Detector	Classifications[a]				Detectivity	Minimum Sample Size	Linearity
	1	2	3	4			
UV	C	Sp	S	N	10^{-10} g/mL	0.5–1 ng	10^5
RI	C	B	U	N	10^{-7} g/mL	1–5 μg	10^4
Fluorometric	C	Sp	S	N	10^{-12} g/mL	10–100 pg	10^5
Electrochemical							
Amperometric	C	Sp	S	N	10^{-10} g/mL	10–500 pg	10^5
Conductometric	C	B	U^b	N	10^{-6} g/mL	10–50 ng	10^3
Transport FID, ECD, TID, PID, MS	MF	Sp	U	D	c	c	10^3

[a] Classifications: 1 = concentration (C) or mass flow rate (MF); 2 = bulk (B) or specific property (SP); 3 = selective (S) or universal (U); 4 = destructive (D) or nondestructive (N).
[b] The conductometric detector is universal for *ions*.
[c] Varies between the GC limits (reported in Chapter 8) and two orders of magnitude less than those limits.

wavelength, the 254 nm mercury line. Both single beam and double beam were available, but the reference beam usually contained a static cell. Their simplicity and intense source give them stability and sensitivity. A typical maximum sensitivity is 0.005 absorbance units full scale (AUFS).

A second wavelength, usually at 280 nm, can be provided in similar instruments by the incorporation of a phosphor that is excited by the mercury line and emits 280 nm.

ASTM has issued a recommended standard practice for testing fixed wavelength detectors in LC; it is ASTM 685-79. In contrast to normal practice, they have defined noise in absorbance units (AU) *per unit cell length*. Otherwise the specifications follow the discussions presented in Chapter 7.

Instruments using monochromators are also available using deuterium lamps, as in conventional instruments. Some can monitor more than one wavelength at a time, and scanning instruments are useful for obtaining a full UV spectrum by stopping the flow for a short time. Unlike the situation in GC, stopping the flow has only a small adverse effect on resolution because of the low diffusion coefficients in liquids. Those instruments that can be used at 190 nm are nearly universal in response, but only a limited number of solvents have sufficient transparency at this wavelength to make them useful in LC.

The new photodiode array instruments[54] are finding extensive use in LC because they can acquire full spectra without stopping the flow. The data are stored on a computer and can be recalled and manipulated to get the maximum amount of information from them. For example, special 3-D (isometric) and 2-D contour plots can be drawn from stored data. Two wavelengths can be ratioed to each other to aid in the detection of unresolved analytes (see Chapter 6) and to characterize analytes qualitatively. Overlapping peaks can be deconvoluted using various mathematical procedures.

Vacancy Chromatography. One way to use a UV detector for analytes that do not absorb is to include a UV absorber as part of the mobile phase. When the non-UV absorbing analytes enter the detector cell, they cause a decrease in the baseline absorbance, appear as negative peaks, and can be used for routine analysis. In effect, they produce a *vacancy*; hence the name for this method of detection.

This effect and similar ones are sometimes responsible for unwanted, unexpected peaks in chromatograms, and then they are called *ghost peaks* or *pseudo peaks*. A recent review of this entire subject[55] calls them *system peaks*, and it is a good source of further details and references.

Refractive Index. Commercial LC refractometers are based on one of two designs, deflection or Fresnel, as shown in Figure 9.28. Both types require reference and sample cells typical of bulk property detectors.

In the deflection type, the beam reflected back from the cells is deflected in proportion to the difference in RI in the sample and reference cells. When no analyte is in the sample cell, the beam is focused on the detector; when an analyte appears in the cell, the beam is deflected and causes a change in signal.

The second type is based on Fresnel's law: the fraction of light reflected at a glass–liquid interface is proportional to the angle of incidence and the relative refractive indices of the substances. For use in LC, the angle of incidence is adjusted so that it is slightly less than the critical angle, and the detector responds to the varying intensity of light striking it. This type can be made with smaller cell volumes than the deflection type, and it is potentially more sensitive. However, it does require two different prisms to cover the entire RI range.

The RI detector requires good temperature control—$\pm 0.001°C$ for maximum sensitivity of 10^{-7} refractive index units. Normally the temperature is controlled to only $\pm 0.01°$, and less than maximum sensitivity is attained. Even this degree of regulation also requires a length of small bore tubing on the inlet side to bring the eluent to detector temperature. As discussed earlier, the volume in this tubing must be minimal to prevent extracolumn peak broadening.

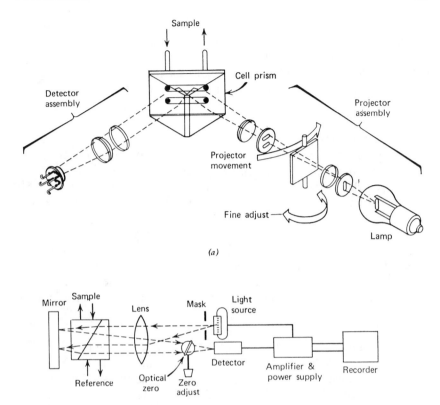

Figure 9.28. Optical diagrams for refractive index detectors for LC: (*a*) Fresnel type; by permission of LDC/Milton Roy; (*b*) deflection type; by permission of Millipore Corporation, Waters Chromatography Division.

In general, the RI detector is not very sensitive, and it cannot be used in gradient elution. Both positive and negative peaks can appear, depending on their refractive indices relative to the mobile phase, and this is considered somewhat disadvantageous. However, it is universal in response, which accounts for its popularity, especially in SEC.

Fluorescence Detector. Typical of fluorescence in general, the fluorescence detector used in LC is about 100 times more sensitive and somewhat more selective than the UV detector. It is this sensitivity that accounts for its popularity and its incorporation into postcolumn reactors. The excitation sources used in LC instruments are as varied as those used in conventional fluorescimeters—from mercury vapor to xenon to quartz halogen to deuterium—but the most interesting designs have used lasers. Lasers have been focused on very small areas, even the end of a capillary column, yielding effective cell volumes as small as 1 nL. These laser-based detectors give increased sensitivity and are very compatible with microbore columns.

Electrochemical Detectors. The only electrochemical detector in current use is amperometric. However, some workers have used the term coulometric for detectors that operate at a high current efficiency and others have used the term polarographic when the electrode is mercury. The acronym LCEC is in common use to represent LC with electrochemical detection.

While electrochemical detectors are extremely popular and useful for some analyses, some analysts find them very difficult to use. However, since they are used in aqueous solutions typical of the popular reverse phase mode, their use continues to expand. They are selective and sensitive, and their very small cell volumes make them ideal for microbore columns.

Other Detectors. The conductivity detector is increasing in popularity because it is usually the detector used in ion chromatographs. The special problems associated with its use were discussed in the section on IC.

The only detector in Table 14 that is not based on an established analytical technique is the transport detector, which uses one of the GC detectors—FID, ECD, PID, or TID—as the measuring device. It consists of a wire, chain, or belt used to deliver the column effluent to the GC detector, removing the volatile mobile phase enroute. Obviously, the samples run on this system must not be volatile, and for them the FID is a universal detector, which is one reason for its use.

Many other detectors are used in LC. *Analytical Chemistry*'s biennial review of column LC[56] contains a very complete discussion of the present state of other detectors, complete with references. The detectors described are atomic absorption, chemiluminescence, conductivity (including photoconductivity), density, dielectric constant, ESR, flame emission, plasma emission, low-angle laser light scattering, evaporative light scattering (mass detector), NMR, optical activity, phosphorescence, photoacoustic, radioactivity, Raman, spectroelectrochemical, streaming current (electrokinetic), thermal energy analyzer, thermal lens calorimetry, and viscometry, plus LC/MS and LC/FTIR, which we will discuss in Chapter 11.

Postcolumn Reactors. Another growing field is the use of postcolumn reactors to produce a species that can be measured by one of the standard detectors, such as UV/visible, fluorescence, or electrochemical. Probably the earliest example of the use of postcolumn reactions was in the determination of amino acids by colorimetry using ninhydrin as the reactant. See the section on derivatization in Chapter 11, as well as the paper in *Analytical Chemistry*,[57] or the book edited by Krull[58] for further details.

Summary. Detector development is an active area of research. While the UV detector is by far the most widely used, there is still a need for a sensitive universal detector. A two-part review has been written by White,[59] and three new books have been published on LC detectors.[60–62]

MOBILE PHASES AND THEIR SELECTION

When each of the various types of LC was discussed earlier in this chapter, the mobile phase was one of the topics included, but a more comprehensive discussion of these liquids, their properties, and their optimal use in LSC and BPC is needed. This section will describe several ways to select the best solvent mixture for a separation. A more comprehensive discussion on the optimization of selectivity has been given by Glajch and Kirkland.[63]

Classification of Solvent Properties

Intermolecular forces were discussed in Chapter 3 and extended to GC stationary phases in Chapter 8. Rohrschneider, followed by McReynolds, investigated the nature of GC stationary phases by using a few common chemicals as probes. Their retention on a given liquid reflected the extent of their interaction with the stationary phase. By choosing probes with selective interactions, they could determine a set of numbers that characterize the liquids under study.

In a similar fashion, Snyder[64] has attempted to characterize LC mobile phases. Using Rohrschneider's data, he has calculated partition coefficients corrected for dispersion interactions and molecular weight effects. As originally defined in Chapter 1, the partition coefficient K is

$$K = \frac{\text{conc. of probe in stationary phase}}{\text{conc. of probe in mobile phase}} \tag{7}$$

To correct for dispersion and molecular weight effects, a new partition coefficient K'' was defined as

$$\log K_g'' = \log K_g' - \log K_v \tag{8}$$

where $K_g' = K_g V_s$ and K_v is a K_g-value for a hypothetical n-alkane with the same molar volume as the probe. Assuming that there are three additive molecular interactions (proton donation, proton acceptance, and strong dipole), the corrected partition coefficients for each effect would add up to a total *polarity index P'*:

$$P' = \log(K_g'')_d + \log(K_g'')_e + \log(K_g'')_n \tag{9}$$

where subscripts d, e, and n refer, respectively, to the three forces.

This polarity index measures the intermolecular attraction between a solute and a solvent, whereas the Hildebrand solubility parameter is defined for pure solvent. For example, ether is not very polar and has a Hildebrand value of 7.4—about the same as hexane, which has a value of 7.3. However, ether can accept protons in the form of hydrogen bonds to its nonbonding electron pairs, and consequently its polarity index is 2.8 compared to 0.1 for hexane.

Snyder's next step was to define three selectivity parameters representing the three forces. To measure proton donation he chose dioxane as the probe; for proton acceptance, ethanol; and for dipolar attraction, nitromethane. Specifically, the selectivity parameter for measuring proton donation, x_d, is

$$x_d = \frac{\log(K_g'')_d}{P'} \tag{10}$$

The other two parameters are defined similarly; the sum of the three parameters is thus normalized to 1. Values for some common solvents are listed in Table 15 (along with the Hildebrand solubility parameters and the Snyder solvent strength parameters).

A three-coordinate plot of these three selectivity parameters is shown in Figure 9.29. Nearly all the solvents studied fit into one of eight groups represented by circles in the figure; only triethylamine and chloroform do not fall into the expected regions. The constituents of each group are listed in Table 16. While the categorization is good, it is not perfect; for example, group 8 contains aromatic hydrocarbons and nitrocompounds that would not seem to have similar attractive forces.

Rohrschneider used five probes and McReynolds ten, but Snyder's

TABLE 15   Polarity Index and Values for Molecular Interactions for Some Solvents

Solvent	P'	χ_e	χ_d	χ_n	Group	ϵ^0	δ
n-Hexane	0.1					0.00	7.3
i-Octane	0.1					0.01	7.0
CCl_4	1.6					0.14	8.6
Toluene	2.4	0.25	0.28	0.47	8	0.22	8.9
Benzene	2.7	0.23	0.32	0.45	8	0.25	9.2
Ethyl ether	2.8	0.53	0.13	0.34	1	0.29	7.4
Methylene chloride	3.1	0.29	0.18	0.53	5	0.32	9.7
i-Propanol	3.9	0.55	0.19	0.27	2	0.63	11.5
n-Propanol	4.0	0.54	0.19	0.27	2	0.63	11.5
Tetrahydrofuran	4.0	0.38	0.20	0.42	3	0.35	9.1
Chloroform	4.1	0.25	0.41	0.33	Close to 8	0.31	9.3
Ethanol	4.3	0.52	0.19	0.29	2	0.68	12.7
Ethyl acetate	4.4	0.34	0.23	0.43	6	0.45	9.6
2-Butanone (MEK)	4.7	0.35	0.22	0.43	6	0.39	9.3
Dioxane	4.8	0.36	0.24	0.40	6	0.43	10.0
Acetone	5.1	0.35	0.23	0.42	6	0.43	9.9
Methanol	5.1	0.48	0.22	0.31	2	0.73	14.4
Acetonitrile	5.8	0.31	0.27	0.42	6	0.50	11.7
Nitromethane	6.0	0.28	0.31	0.40	7	0.49	12.6
Water	10.2	0.37	0.37	0.25	8	Large	21

Source: Reproduced from the *Journal of Chromatographic Science* by permission of Preston Publications, Inc.

approach has shown that three probes are adequate. Specifically, the addition of toluene and 2-butanone, which would give five probes very similar to Rohrschneider's original five, has been found to provide little additional information. The probes used by the various workers are compared in Table 17.

To use the selectivity parameters in selecting solvent mixtures for LC, Snyder assumes additivity in proportion to the volume fractions of the solvents:

$$P' = \left(\frac{V_A}{V_B}\right) P_A + \left(\frac{V_B}{V_A}\right) P_B \qquad (11)$$

The solvents chosen, A and B, should be from two of the eight solvent groups and be as different as possible. Since it has been shown that three solvents are better than two, the ideal ternary mixture for normal phase LC should have three solvents chosen from the three apices of the triangle.

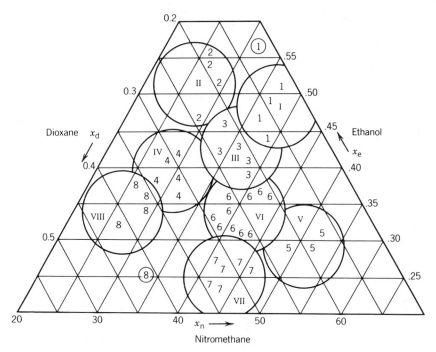

Figure 9.29. Selectivity grouping of solvents in Table 15. Reproduced from the *Journal of Chromatographic Science* by permission of Preston Publications, Inc.

TABLE 16 Classification of Solvents in Figure 9.29

Group	Solvents
1	Aliphatic ethers, tetramethylguanidine, hexamethyl phosphoric acid amide, (trialkylamines)[a]
2	Aliphatic alcohols
3	Pyridine derivatives, tetrahydrofuran, amides (except formamide), glycol ethers, sulfoxides
4	Glycols, benzyl alcohol, acetic acid, formamide
5	Methylene chloride, ethylene chloride
6	(a)[b] Tricresyl phosphate, aliphatic ketones and esters, polyethers, dioxane
	(b)[b] Sulfones, nitriles, propylene carbonate
7	Aromatic hydrocarbons, halo-substituted aromatic hydrocarbons, nitro compounds, aromatic ethers
8	Fluoroalkanols, *m*-cresol, water (chloroform)[c]

[a] Somewhat more basic than other group 1 solvents.
[b] This group is rather broad and can be subdivided as indicated into groups 6a and 6b; however, normally there is no point to this in practical usage of the present scheme.
[c] Somewhat less basic than other group 8 solvents.

TABLE 17   Probes Used in Determining Polarity

Rohrschneider	McReynolds	Snyder
Benzene	Benzene	
Ethanol	n-Butanol	Ethanol
2-Butanone	2-Pentanone	
Nitromethane	Nitropropane	Nitromethane
Pyridine	Pyridine	
	2-Methyl-2-pentanol	
	Iodobutane	
	2-Octyne	
	1,4-Dioxane	1,4-Dioxane
	cis-Hydrindane	

Snyder recommended ethyl ether from group 1, methylene chloride from group 5, and chloroform, which is usually included in group 8, because these three are widely separated on the plot (Figure 9.29).

Before applying this strategy to a normal phase system, we need to consider some additional procedures that have become part of the optimization process.

Overlapping Resolution Mapping

In 1980 Kirkland and co-workers[65] applied the strategy developed by Snyder in conjunction with a mixture-design statistical technique. They arrived at a new method of data analysis that they called *overlapping resolution mapping* (ORM). The steps in the method are: (1) selection of the three ideal solvents à la Snyder; (2) chromatographing the sample mixture up to ten times using different mixtures of the three solvents chosen statistically; (3) preparation of resolution maps for each pair of analytes in the mixture; and (4) overlapping these maps to obtain an ORP that designates the ideal mixture(s) to be used for optimal separation.

For NPLC, the three solvents selected were the same as those recommended by Snyder, except that methyl-*t*-butyl ether is substituted for ethyl ether. The (nonpolar) carrier solvent recommended is hexane, and the stationary phase is silica.

For RPLC, the three solvents chosen were methanol from group 2, acetonitrile (ACN) from group 6, and tetrahydrofuran (THF) from group 3. These three solvents are not as widely separated from each other in the diagram as those for normal phase are, but they are sufficiently different to produce good separations. The (polar) carrier solvent used was water, and the stationary phase was a C_8 bonded phase. The procedure will be illustrated first with a reverse phase separation of nine substituted naphthalenes.

First, the concentration of each of the three solvents in water must be optimized to produce reasonable retention times. As we have seen, the ideal situation exists when the partition ratios are in the range of about 1.5 to 5, and slightly higher and lower values can be considered. For this example, the ideal concentrations were:

Mobile phase A: 63% methanol in water.
Mobile phase B: 52% ACN in water.
Mobile phase C: 39% THF in water.

The incorporation of these three solvents into a statistical design is shown in Figure 9.30. The ten mixtures shown include the three solvents (points 1, 2, and 3 at the apices), 50/50 mixtures of pairs of solvents (points 4, 5 and 6), a 33/33/33 mixture of all three (point 7), and 16/16/67 mixtures of all three (points 8, 9, and 10). The data for the first seven runs in this study is shown in Figure 9.31, where it can be seen that some unexpected variations occur but there are no crossovers in peak elution order.

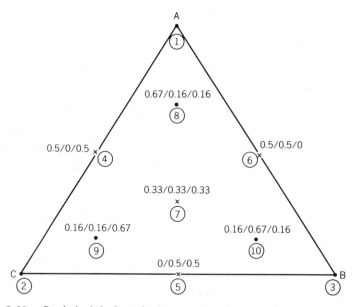

Figure 9.30.   Statistical design of mixtures of ternary mobile phases. Reprinted with permission from ref. 65.

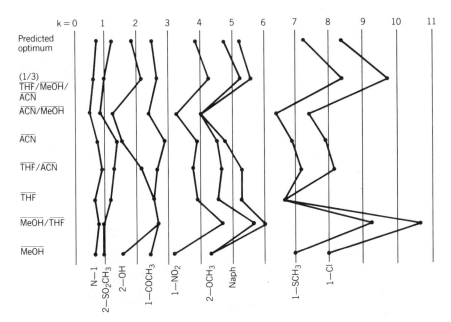

Figure 9.31. Solvent selectivity data summary for eight mobile phases. Reprinted with permission from ref. 65.

Next, a computer-generated resolution map is made for each (adjacent) pair of peaks using the resolution values from the runs (only seven in this example out of the possible ten). The contour map for peaks 8 and 9 is shown in Figure 9.32, which also gives the seven resolution values found in the seven runs. An arbitrary minimal resolution (e.g., 1.5) is chosen, and the portion of the map giving a resolution equal to or greater than that value is shown as clear space in Figure 9.33. Three pairs of peaks (4/5, 5/6, and 7/8) are resolved by all solvents and can be disregarded. The other five maps are overlapped to see what areas of good resolution they have in common (Figure 9.34). Any solvent mixture falling in the clear region labelled A will give a resolution of at least 1.5 for all eight pairs of peaks. The point marked X in the figure is an optimal mixture chosen by another method described in the same paper.

If the elution order changes, additional maps must be prepared to consider the original order plus all the additional new pairs, but the ORM method can handle such a situation. For a complicated mixture, the procedure may become too laborious or the individual components may not be available for preparing a standard test mixture. Perhaps some easy-to-separate components can be eliminated from the screening, and selec-

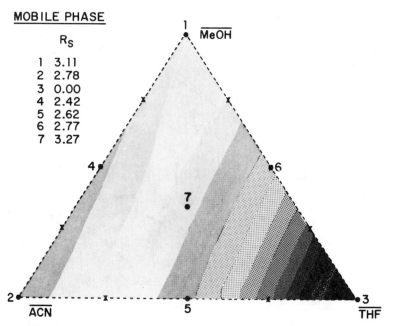

Figure 9.32. Resolution map for peaks 8 and 9. Reprinted with permission from ref. 65.

tive detectors can be used to establish the identity of eluting components to detect easily changes in elution order.

The situation in LSC is perceived to be slightly different because of the competition that exists between the solvents and the analytes for active sites on the solid stationary phase surface—so-called *localization effects*. Snyder has been most active in describing the theoretical basis for LSC, and in 1982 he joined with Glajch and Kirkland to describe the application of the ORM technique in LSC.[66] Their choice of solvents for LSC is slightly different from those proposed earlier because of the localization effect. Thus, they recommend: (1) a nonlocalizing solvent (methylene chloride, the same as earlier); (2) a basic, localizing solvent (methyl-*t*-butyl ether, the same as earlier); and (3) another localizing solvent that is not basic (acetonitrile, a new recommendation). Hexane was chosen as the nonpolar carrier solvent as before, although 1,1,2-trifluorotrichloroethane is also a possibility.

The optimization procedure is similar to that used in reverse phase. The concentration of methylene chloride in hexane is adjusted to get partition ratios in the range of 1 to 10. The solvent strength ϵ^0 of this solution is calculated and used in preparing the other solvent mixtures to the same value. Consequently, all mixtures used in this process will have the same

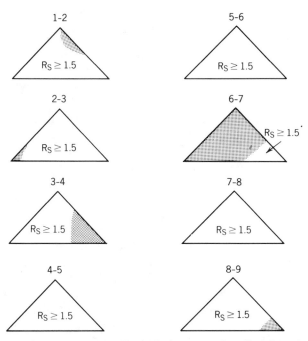

Figure 9.33. Simplified ($R_s > 1.5$) resolution maps for all eight pairs of peaks. Reprinted with permission from ref. 65.

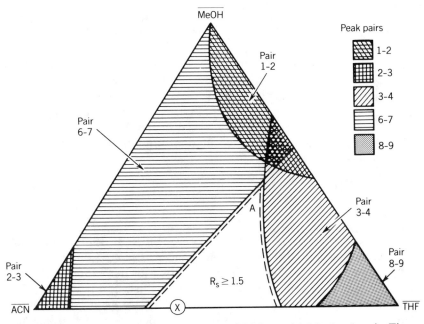

Figure 9.34. Overlapping resolution map (ORM) combining the data in Figure 9.33. Reprinted with permission from ref. 65.

ϵ^0 values. (Note: it is anticipated that sometimes additional methylene chloride may be needed in the ACN mixtures in order to get miscibility.) The seven solvent mixtures are prepared and the sample is run in each; contour maps are made for adjacent pairs of peaks; and, finally, an ORM is prepared and the optimum region is identified.

Other workers have successfully applied these principles to the optimization of their separations, and their papers attest to the value of this method: Antle[67] has separated steroids by normal phase on an amine bonded phase, and Lehrer[68] has separated phenols and cresols by reverse phase on a C_8 bonded phase. Both provide interesting chromatograms.

Phase Selection Diagrams

A number of other methods has been used to optimize ternary solvent systems, many of them[69-71] similar to the "window diagram" used in GC. Only one more will be described briefly here. The authors, Schoenmakers et al.,[72] have prepared what they call phase selection diagrams for several reverse phase LC separations. The solvents used for their mixtures are the same ones recommended by Snyder: methanol, ACN, and THF.

First, they make a gradient run with one of the solvents (usually methanol) in water. From that run they choose the best isocratic conditions and run the sample with it. Then they choose a binary mixture of one of the other solvents in water, prepare a mixture that is iso-eluotropic (same solvent strength, using Snyder's solvent strength parameter), and rerun the sample. In one example, the first binary mixture was 50% methanol, 50% water, and the other mixture was 32% THF and 68% water. The ln k values for these two runs are plotted on the y axis of Figure 9.35. It can be seen that the elution order in the first case was 1 (and 2 unresolved), 3, 4, 5, 6 and in the second was 1, 3, 2, 6, 5, 4. The two points for each peak are joined by straight lines (dashed in the figure), and the phase selection diagram is drawn by calculation of an *optimization criterion*, $\prod R_s$, defined as:

$$\prod R_s = \prod_{i=1}^{n-1} \frac{k_{i+1} - k_i}{k_{i+1} + k_i + 2} \tag{12}$$

where n is the number of peaks in the chromatogram. Equation (12) can be derived from Eq. (18) in Chapter 4, and it is plotted on an arbitrary scale on the right in the figure. It predicts that, for this example, the best separation will occur at its maximum value, which occurs at a (ternary) mixture of 10% methanol, 25% THF, and 65% water, as shown in the figure. In this case, the predicted separation does occur. If it does not,

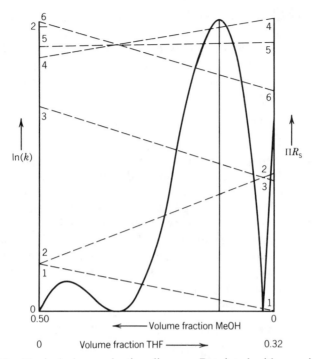

Figure 9.35. Typical phase selection diagram. Reprinted with permission from ref. 72.

a new revised diagram is prepared using the data from the three runs, and a new optimum mixture is predicted. Thus, with only a few trials, a satisfactory separation can be achieved.

SAMPLE CLEAN-UP AND PREPARATIVE ISOLATION

LC is often used to clean up samples to remove unwanted matrix interferences. Similarly, the isolation of a newly synthesized compound from its reaction mixture or a newly identified chemical from a natural product can be achieved with large columns that allow the separation of large amounts of material. This section discusses some of the factors to be considered in these types of separations, which are usually known as *preparative*, or *prep* methods. Most of the discussion will be on low pressure methods, but, of course, the principles also apply to high pressure operation.

There are no definite divisions between low and high pressure and between analytical and preparative sample sizes. By low pressure LC, we will mean operation under gravity flow up to several hundred kPa of pressure. Prep LC will be loosely defined as sample sizes ranging from a few hundred mg up to several grams. Typically prep separations are performed with sample sizes ≥ 1 mg of sample per gram of column packing.

Sample Clean-Up

A number of manufacturers now supply short LC columns to be used under gravity flow or under the flow induced by vacuum, a centrifuge, or a syringe. Some of the trade names used are Sample Enrichment and Purification (SEP-PAC), PreSep Extraction, Bond-Elute, and solid-phase extraction. The columns are a few inches long and can be packed with any of the stationary phases used in LC, although silica annd bonded phases are most common. The separations take a few minutes, and the eluent is collected for further analysis. The inexpensive cartridges are disposed of after use. Majors[73] has compared the various devices commercially available in 1986.

Low Pressure Prep LC

The type of prep separations included in this section vary widely; they are not instrumental methods, although some pumps, detectors, injectors, and fraction collectors may be used. They are simple, similar to the original work by Tswett, and sometimes called *classical LC*. The paper by Crane, Zief, and Horvath[74] is typical.

The stationary phase most often used is silica, but other solids are also listed in Table 18. Newer bonded phases can also be used, but they are more expensive. Relatively large particle sizes are needed to keep the

TABLE 18   Solids Used as Stationary Phases in LSC (in Order of Increasing Activity)

1. Sucrose	9. Calcium phosphate
2. Starch	10. Magnesium carbonate
3. Inulin	11. Magnesium oxide
4. Magnesium citrate	12. Silica gel
5. Talc	13. Magnesium silicate (Florisil)
6. Sodium carbonate	14. Alumina
7. Potassium carbonate	15. Charcoal, activated
8. Calcium carbonate	16. Fuller's earth (kaolin-type clay)

pressure drop low and facilitate easy dry packing. The range of particle sizes is often large for the sake of economy if high efficiency is not needed.

The columns are often made of glass and consequently can be used only at low pressures. Compared to analytical columns, they are short and fat with inside diameters around 1 to 4 cm and lengths of 20 to 100 cm. In all prep columns, higher efficiency is attained if the bottom of the column is rounded, and that configuration is easily achieved with glass. The volume of the column, and hence its sample capacity, goes up with the square of the column diameter. The amount of stationary phase needed to pack the column increases by the same proportion, which is the reason for using low cost packings. Some typical operating conditions are given in Table 19.

The mobile phase should have the requirements discussed in the LSC section, but in particular it should be inexpensive and volatile to facilitate sample recovery. If silica gel is the stationary phase, the amount of water in the system becomes critical, and care must be exercised since inexpensive solvents may be wet. Inexpensive solvents may also contain less volatile impurities that will be concentrated when the solvent is evaporated, causing contamination of the isolated sample.

The mobile phase can be allowed to flow under the force of gravity, a low pressure pump can be used, or compressed gas can be used to pressurize a solvent reservoir. The linear velocity should be about one-third of that used in analytical columns. The sample can be applied to the top of the column with a microsyringe or pipet using the stop-flow method, or an inexpensive, low-pressure valve can be used. The eluent is usually collected in separate tubes using an automated fraction collector. Inexpensive UV detectors with large solvent volumes are available, or flow cells can be fitted to conventional UV/visible instruments.

TABLE 19   Typical Operating Conditions for
Low Pressure Preparative LC

Column i.d.	1–4 cm
Column length	20–100 cm
Weight of column packing	50–500 g
Particle size	>40 μm
Operating pressure	10 to 300 kPa
Flow rate	2 to 20 mL/min
Linear velocity	0.1 to 1.0 mm/sec
Plate number	200 to 2000
Sample size	0.1 to 10 g
Sample volume	0.5 to 20 mL

Since the samples run in prep LC are often very crude, it is to be expected that the column packing will become contaminated with chemicals that did not elute. Cleaning can be accomplished with polar solvents like propanol or ethyl acetate (for normal phase systems), or the packing can simply be discarded. The latter alternative may be cheaper, considering the cost of solvents and waste disposal.

Often TLC is used for fast screening of mobile phases, since the stationary phases used in prep LC are also available for TLC. The TLC R_f value should be equal to or less than 0.3. In prep work, the separation is usually optimized so that the column can then be run with a sample overload and still produce adequate separations. Overloading will cause the plate number and partition ratio to decrease, and chromatograms produced this way are not pretty to look at.

Flash Chromatography. In 1978, Still, Kahn, and Mitra[75] published a "quick-and-dirty" prep method that has become very popular and even has its own name—*flash chromatography.* They used short, fat columns (1–5 cm i.d. × 18 in.), 40–65 μm silica, and low viscosity solvents (e.g., ethyl acetate/petroleum ether mixtures). The exact solvent mixture is selected by screening the sample via TLC to give an R_f value of about 0.35. However, they found that when the proportion of the polar component was small, it was better to use about half as much of it for the column separation as was used for TLC.

Basically, the procedure is as follows. The column is partially filled (5–6 in.) with the silica gel, and solvent is added to fill it. Compressed air is used to flush the solvent quickly through the packing, compressing it and removing all air from it. The sample is then added to the top of the column, and the column is again filled with solvent. The pressure is adjusted to get a flow of 2 in./min, and the separation is accomplished in 5 to 10 min.

Dry Column LC. Another variation of low pressure LC is performed on a column of dry packing[76] similar to conventional TLC. The column is a Nylon tube with a hole in the bottom. When the mobile phase reaches the end of the column, flow is stopped, so the sample is not eluted. Rather, the column is sliced, and the components are recovered from the packing after their locations are established by fluorescence or staining reagents. This column method can handle larger samples than prep TLC.

High Performance Prep LC

For difficult separations, high performance columns are required, which means small particle columns and instrumentation. The use of an instrument will also improve the speed of analysis and offers the possibility of automatic operation.

One possibility is to scale up an analytical instrument for prep work. High efficiency prep columns are available for that purpose, but column diameters only 2 or 3 times analytical size are usable on analytical instruments. The limitations are the pump and the detector. A column 3 times as large as the analytical column will require 9 times as much flow to maintain the same linear velocity, and this flow rate may approach the limit of the pump. The detector must be capable of handling these higher flows and sample sizes without overloading. As a compromise, a semiprep configuration is often used, and the sample is run repeatedly to get sufficient sample isolated.

Until recently, one company dominated the market for preparative units. The success of prep LC has encouraged other manufacturers to enter the market, and a recent review[77] lists eight commercially available units, three of which can be operated automatically and unattended. These instruments can also accommodate larger columns and hence larger samples. Capacities up to 750 g and flows of 5 L/min are claimed. Columns are relatively expensive and overload easily. In addition to normal slurry packing techniques, radial and axial compression devices are used in some of them to improve performance.

The refractive index detector is popular in prep units because it is universal and less sensitive. UV detectors are often too sensitive, and short path length cells are needed. Alternatively, the UV detector can be detuned to a less sensitive wavelength, or the effluent stream can be split, sending only part of it through the detector. A unique UV detector uses a thin film instead of a cell and claims typical path lengths around 0.2 mm and flows up to 400 mL/min.[78]

The maximum sample size that can be injected without a loss in resolution can be calculated with Eq. (13)[79]:

$$V_{inj} = V_M\left[(k_B - k_A) - \left(\frac{2}{\sqrt{n}}\right)(2 + k_A + k_B)\right] \qquad (13)$$

Since the dead volume V_M is rather large in prep columns, the size of the sample volume can also be quite large and may require a pump for filling the sample loops in the sample valve. For maximum efficiency, the sample should be spread out over the entire column, and a study of this effect has recommended special inlet configurations.[80]

For further information, including a bibliography of 135 references, see reference 77, or the biennial review in *Analytical Chemistry*,[56] or a four-year review covering 1980–1984.[81]

SPECIAL TOPICS

There are a few topics that need to be introduced here even though space does not permit them to be thoroughly discussed. References are provided for further information.

Pseudophase or Micellar LC

Aqueous micellar solutions were first used as mobile phases in LC in 1980. A micellar solution is one that contains a surfactant at a concentration above the critical micelle concentration, about 10^{-2} M. The nature of micelles and their use in analytical chemistry, including chromatography, has been recently described.[82]

For LC, they have been used with nonpolar bonded phases and are typical of reverse phase separations as we have discussed them. They have also been called *pseudophase LC*; at present they seem to be unique enough to warrant such special names, but it may turn out that they are just special forms of reverse phase LC.

The surfactants that have been used are the anionic dodecylsulfate, cationic quarternaries such as hexadecyltrimethylammonium ion, and, more recently,[83] nonionics such as Triton X-100 and Brij-35. The solvent is water or water–organic mixtures. The special properties of the mobile phases are claimed to provide unique selectivities and have been shown to be capable of handling direct injection of blood serum samples for therapeutic drug monitoring.[84] Gradients can be run, and no equilibration is necessary to return them to the original conditions, thus saving time. See the introduction to reference 83 for a recent literature summary.

Affinity Chromatography

Affinity chromatography is characterized by its unique stationary phase, which has a specific bioactive ligand bonded onto a solid support. These bonded phases are somewhat similar to the other bonded phases discussed earlier, but their specificity sets them apart. All applications have been with biomolecules that show the necessary specific interactions. Typical bonded ligands and their specificities are listed in Table 20. The "active" component of a sample is the only one attracted to the stationary phase, and the others are washed off; then the eluent is changed to desorb the single analyte. While there are other less specific applications, the highly specific ones are not subject to the normal chromatographic equilibrium processes, and the chromatographic theory we have presented does not

TABLE 20   General Ligands and Their Specificities[a]

Ligand	Specificity
Cibacron Blue F3G-A dye, derivatives of AMP, NADH, and NADPH	Certain dehydrogenases via binding at the nucleotide binding site
Concanavalin A, lentil lectin, wheat germ lectin	Polysaccharides, glycoproteins, glycolipids, and membrane proteins containing sugar residues of certain configurations
Soybean trypsin inhibitor, methyl esters of various amino acids, D-amino acids	Various proteases
Phenylboronic acid	Glycosylated hemoglobins, sugars, nucleic acids, and other cis-diol-containing substances
Protein A	Many immunoglobulin classes and subclasses via binding to the F_c region
DNA, RNA, nucleosides, nucleotides	Nucleases, polymerases, nucleic acid

[a] From Walters,[85]

apply. Highly specific columns can be as short as 1 cm and separation times as short as 1 min.

Three thorough reviews[85-87] and at least six books[88-93] are available for further study.

Internal Surface Reverse Phase Supports

In 1985, Hagestam and Pinkerton[94] published a report on a new type of stationary phase they synthesized. They called it an *internal surface reverse phase* (ISRP) support, and it is also known as a *Pinkerton* column. The patented idea has been licensed exclusively to Regis Chemical Company.

The concept behind the invention is shown in Figure 9.36. It operates with two mechanisms—size exclusion and reverse phase bonded sorption. The purpose is to facilitate the injection of plasma samples without prior clean up to remove proteins that normally clog a reverse phase column. The pore size of the stationary phase is small enough that the large protein molecules cannot enter and the outer surfaces to which they are exposed do not retain them at all, so they are eluted off the column

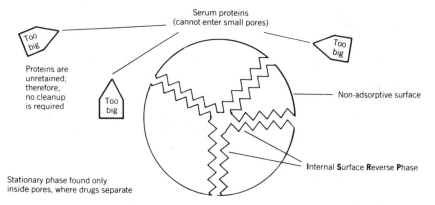

Figure 9.36. Representation of internal surface reverse phase (ISRP) type support. Courtesy of Regis Chemical.

quickly. The smaller analyte molecules penetrate the pores where they experience interactions with the bonded phase and are retained and chromatographed. The columns are too new to have been thoroughly evaluated, but the approach is novel.

Instrumental Topics

A few special instrument configurations are worthy of a brief note. In cases where a single pass through a column has not produced adequate resolution, the eluent from the column can be passed through the column again. This process has become known as *recycle* chromatography and has been most often used in size exclusion and prep chromatography. As a minimum, a valve is required to divert the eluent back to the pump where it replaces the mobile phase. Often, however, this process also includes some *heartcutting*, especially in prep LC. For further details see the review in reference 95.

As in GC, it is becoming increasingly popular to include the possibility of using successive columns in LC instruments. Examples are usually classed as "multidimensional," and some are listed in the biennial *Analytical Chemistry* review.[56]

Finally, it should be noted that LC is reliable enough to be run automatically with an autosampler, and a large number of laboratories perform their analyses that way. The increasing availability of robots extend the possibilities, especially in sample preparation, and they will undoubtedly increase in importance.

TABLE 21 Evaluation of Four Commercial 15-cm C_{18} Columns at a Reduced Flow of Approximately 5^a

Parameters	Column			
	A	B	C	D
ν	4.9	5.2	4.7	4.8
F^b (mL/min)	0.50	0.60	0.60	0.55
k, toluene	3.8	3.3	2.1	2.5
n^c	13,200	11,900	12,100	10,600
H (μm)	11.3	12.6	12.4	14.1
h	1.9	2.5	2.5	3.5
Φ	1090	630	690	510
ΔP (bar)	34	36	35	52
E	3900	4000	4300	6300

^{a} From Pauls and McCoy.[96] Reproduced from the *Journal of Chromatographic Science* by permission of Preston Publications, Inc.
^{b} Volumetric flow rate.
^{c} Calculated employing width at half height.

EVALUATION

The *quality* of LC columns and LC separations can be measured in several ways. A recent paper[96] summarizes some testing procedures, and Table 21 is taken from that publication. Four commercial columns are compared using the column parameters we have defined and described: n, H, h, and Φ. Although variations exist between the columns, the values presented provide convenient comparisons against which one can judge his/her own column's performance. It is highly recommended that any new or repacked column be tested to evaluate its performance, and that it be periodically reevaluated to determine when it needs to be repacked or dis-

TABLE 22 Evaluation of LC

Advantages	Disadvantages
1. Applicable to nonvolatile samples	1. Slower than GC
2. Efficient, selective, and widely applicable	2. Experimentally more complex than GC
3. Can be quantitated	
4. Requires only a small sample	
5. Nondestructive	

carded. Reference 96 should be consulted for additional details on column evaluation.

Table 22 contains a summary of the advantages and disadvantages of LC in columns. Although LC is compared to GC, the two techniques should be considered to be complementary, not competitive—each has its own special advantages and applications.

REFERENCES

1. R. E. Majors, *LC Mag.* **1985,** *3,* 774.
2. B. W. Sands, Y. S. Kim, and J. L. Bass, *J. Chromatogr.* **1986,** *360,* 353.
3. R. E. Majors, *Anal. Chem.* **1972,** *44,* 1722.
4. J. M. Anderson, *J. Chromatogr. Sci.* **1984,** *22,* 343.
5. K. K. Unger, *Porous Silica,* Elsevier, New York, 1979.
6. R. P. W. Scott, *Adv. Chromatogr. N. Y.* **1982,** *20,* 167.
7. L. R. Snyder, *Principles of Adsorption Chromatography,* Dekker, New York, 1968.
8. R. W. Yost and R. D. Conlon, *Chromatogr. Newsletter (Perkin-Elmer),* **1972,** *1*(1), 5.
9. D. L. Saunders, *J. Chromatogr. Sci.* **1977,** *15,* 372.
10. P. Jandera and J. Churacek, *Gradient Elution in Column Liquid Chromatography,* Elsevier, New York, 1985.
11. L. R. Snyder and D. L. Saunders, *J. Chromatogr. Sci.* **1969,** *7,* 195.
12. J. F. K. Huber, C. A. M. Meyers, and J. A. R. J. Hulsman, *Anal. Chem.* **1972,** *44,* 111; J. F. K. Huber, E. T. Alderliste, H. Harren, and H. Poppe, *Anal. Chem.* **1973,** *45,* 1337.
13. J. N. Driscoll and I. S. Krull, *Am. Lab.* **1983,** *15*(5), 42.
14. R. K. Gilpin, *J. Chromatogr. Sci.* **1984,** *22,* 371.
15. S. D. Fazio, S. A. Tomellini, H. Shih-Hsien, J. B. Crowther, T. V. Raglione, F. R. Floyd, and R. A. Hartwick, *Anal. Chem.* **1985,** *57,* 1559.
16. C. Horvath and W. Melander, *J. Chromatogr. Sci.* **1977,** *15,* 393.
17. D. C. Locke, *J. Chromatogr. Sci.* **1974,** *12,* 433.
18. R. K. Gilpin, *Anal. Chem.* **1985,** *57,* 1465A.
19. S. Hjerten, *Adv. Chromatogr. N. Y.* **1981,** *19,* 59.
20. S. Moore and W. H. Stein, *J. Biol. Chem.* **1951,** *192,* 663.
21. K. A. Kraus and G. E. Moore, *J. Am. Chem. Soc.* **1953,** *75,* 1460.
22. F. M. Rabel, *Adv. Chromatogr. N. Y.* **1979,** *17,* 53.
23. R. M. Baum, *Chem. Eng. News* **1985,** May 20, p. 9.
24. H. Small, *Anal. Chem.* **1983,** *55,* 235A.

25. H. Small, T. S. Stevens, and W. C. Bauman, *Anal. Chem.* **1975**, *47*, 1801.
26. T. S. Stevens, G. L. Jewett, and R. A. Bredeweg, *Anal. Chem.* **1982**, *54*, 1206.
27. T. S. Stevens and M. A. Langhorst, *Anal. Chem.* **1982**, *54*, 950.
28. H. Small and T. E. Miller, *Anal. Chem.* **1982**, *54*, 462.
29. S. Eksborg, P. O. Lagerstrom, R. Modin, and G. Schill, *J. Chromatogr.* **1973**, *83*, 99; S. Eksborg and G. Schill, *Anal. Chem.* **1973**, *45*, 2092.
30. J. H. Knox and G. R. Laird, *J. Chromatogr.* **1976**, *122*, 17.
31. B. A. Bidlingmeyer, *J. Chromatogr. Sci.* **1980**, *18*, 525.
32. M. T. W. Hearn, *Adv. Chromatogr. N. Y.* **1980**, *18*, 59; M. T. W. Hearn (ed.), *Ion-Pair Chromatography*, Dekker, New York, 1985.
33. F. M. Rabel, *J. Chromatogr. Sci.* **1980**, *18*, 394.
34. J. D. Navratil and H. F. Walton, *Am. Lab.* **1976**, *8*(1), 69.
35. V. A. Davankov and A. V. Semechkin, *J. Chromatogr.* **1977**, *141*, 313.
36. D. D. Bly, *Science* **1970**, *168*, 527.
37. J. Cazes, *J. Chem. Educ.* **1970**, *47*, A461 and A505.
38. J. Cazes, *J. Chem. Educ.* **1966**, *43*, A567 and A625.
39. W. W. Yau, J. J. Kirkland, and D. D. Bly, *Modern Size Exclusion Chromatography*, Wiley, New York, 1979.
40. H. M. McNair, *J. Chromatogr. Sci.* **1984**, *22*, 521.
41. J. A. Riddick and W. B. Bunger in *Technique of Chemistry*, 3d ed., Vol. 2, *Organic Solvents*, A. Weissberger (ed.), Wiley, New York, 1970.
42. P. Jandera and J. Churacek, *Gradient Elution in Column Liquid Chromatography*, Elsevier, Amsterdam, 1985.
43. C. Liteanu and S. Gocan, *Gradient Elution Chromatography*, Wiley, New York, 1974.
44. L. R. Snyder, J. W. Dolan, and J. R. Gant, *J. Chromatogr.* **1979**, *165*, 3 and 31.
45. J. W. Dolan, *LC Mag.* **1985**, *3*, 1050.
46. S. van der Wal, *LC Mag.* **1985**, *3*, 488.
47. P. Kucera (ed.), *Microcolumn High-Performance Liquid Chromatography*, Elsevier, Amsterdam, 1984.
48. R. P. W. Scott (ed.), *Small Bore Liquid Chromatography Columns*, Wiley, New York, 1984.
49. M. Novotny and D. Ishii (eds.), *Microcolumn Separation Methods*, Elsevier, Amsterdam, 1985.
50. E. J. Caliguri, P. Capella, L. Bottari, and I. N. Mefford, *Anal. Chem.* **1985**, *57*, 2423.
51. R. P. W. Scott, *Adv. Chromatogr. N. Y.* **1983**, *22*, 247.
52. M. Novotny, *LC Mag.* **1985**, *3*, 876.

53. S. A. Borman, *Anal. Chem.* **1982**, *54*, 327A.

54. D. G. Jones, *Anal. Chem.* **1985**, *57*, 1057A and 1207A.

55. S. Levin and E. Grushka, *Anal. Chem.* **1986**, *58*, 1602.

56. H. G. Barth, W. E. Barber, C. H. Lochmuller, R. E. Majors, and F. E. Regnier, *Anal. Chem.* **1986**, *58*, 211R.

57. R. W. Frei, H. Jansen, and U. A. Th. Brinkman, *Anal. Chem.* **1985**, *57*, 1529A.

58. I. S. Krull (ed.), *Reaction Detection in Liquid Chromatography*, Dekker, New York, 1986.

59. P. C. White, *Analyst* **1984**, *109*, 677 and 973.

60. T. M. Vickrey (ed.), *Liquid Chromatography Detectors*, Dekker, New York, 1983.

61. R. P. W. Scott, *Liquid Chromatography Detectors*, 2d ed., Elsevier, Amsterdam, 1986.

62. E. S. Yeung, *Detectors for Liquid Chromatography*, Wiley-Interscience, New York, 1986.

63. J. L. Glajch and J. J. Kirkland, *Anal. Chem.* **1983**, *55*, 319A.

64. L. R. Snyder, *J. Chromatogr. Sci.* **1978**, *16*, 223.

65. J. L. Glajch, J. J. Kirkland, K. M. Squire, and J. M. Minor, *J. Chromatogr.* **1980**, *199*, 57.

66. J. L. Glajch, J. J. Kirkland, and L. R. Snyder, *J. Chromatogr.* **1982**, *238*, 269.

67. P. E. Antle, *Chromatographia* **1982**, *15*, 277.

68. R. Lehrer, *Am. Lab.* **1981**, *13*(10), 113.

69. R. J. Laub, *Am. Lab.* **1981**, *13*(3), 47.

70. S. N. Deming and M. L. H. Turoff, *Anal. Chem.* **1978**, *50*, 546.

71. B. Sachok, R. C. Kong, and S. N. Deming, *J. Chromatogr.* **1980**, *199*, 317.

72. P. J. Schoenmakers, A. C. J. H. Drouen, H. A. H. Billiet, and L. de Galan, *Chromatographia* **1982**, *15*, 688.

73. R. E. Majors, *LC/GC Mag.* **1986**, *4*, 972.

74. L. J. Crane, M. Zief, and J. Horvath, *Am. Lab.* **1981**, *13*(5), 128.

75. W. C. Still, M. Kahn, and A. Mitra, *J. Org. Chem.* **1978**, *43*, 2923.

76. B. P. Engelbrecht and K. A. Weinberger, *Am. Lab.* **1977**, *9*(5), 71.

77. R. Sitrin, P. DePhillips, J. Dingerdissen, K. Erhard, and J. Filan, *LC/GC Mag.* **1986**, *4*, 530.

78. J. M. Miller and R. Strusz, *Am. Lab.* **1980**, *12*(1), 29.

79. R. P. W. Scott and P. Kucera, *J. Chromatogr.* **1976**, *119*, 467.

80. A. W. J. De Jong, H. Poppe, and J. C. Kraak, *J. Chromatogr.* **1978**, *148*, 127.

81. M. Verzele and C. Dewaele, *LC Mag.* **1985**, *3*, 22.

82. L. J. Kline Love, J. G. Habarta, and J. G. Dorsey, *Anal. Chem.* **1984,** *56,* 1132A.

83. M. F. Borgerding and W. L. Hinze, *Anal. Chem.* **1985,** *57,* 2183.

84. F. J. DeLuccia, M. Arunyanart, and L. J. Kline Love, *Anal. Chem.* **1985,** *57,* 1564.

85. R. R. Walters, *Anal. Chem.* **1985,** *57,* 1099A.

86. I. Parikh and P. Cuatrecasas, *Chem. Eng. News* **1985,** *63*(34), 17.

87. P.-O. Larsson, M. Glad, L. Hansson, M.-O. Mansson, S. Ohlson, and K. Mosbach, *Adv. Chromatogr., N.Y.* **1983,** *21,* 41.

88. C. R. Lowe and P. D. G. Dean, *Affinity Chromatography*, Wiley, New York, 1974.

89. J. Turkova, *Affinity Chromatography*, Elsevier, Amsterdam, 1978.

90. W. H. Scouten, *Affinity Chromatography*, Wiley, New York, 1981.

91. I. Chaiken, M. Wilchek, and I. Parikh (eds.), *Affinity Chromatography and Biological Recognition*, Academic Press, New York, 1984.

92. H. Schott, *Affinity Chromatography: Template Chromatography of Nucleic Acids and Proteins*, Dekker, New York, 1985.

93. P. Mohr and K. Pommerening, *Affinity Chromatography*, Dekker, New York, 1985.

94. I. H. Hagestam and T. C. Pinkerton, *Anal. Chem.* **1985,** *57,* 1757.

95. M. Martin, F. Verillon, C. Eon, and G. Guiochon, *J. Chromatogr.* **1976,** *125,* 17.

96. R. E. Pauls and R. W. McCoy, *J. Chromatogr. Sci.* **1986,** *24,* 66.

LIQUID CHROMATOGRAPHY
ON PLANE SURFACES

10

There are two popular LC techniques in which the stationary bed is supported on a planar surface rather than in a column: paper chromatography (PC) and thin-layer chromatography (TLC). PC preceded TLC by some 10 to 15 years, and a large number of excellent separations were devised for it. But beginning about 1956, it was found that TLC could also be used for most of these separations and that it was faster, more reproducible, more versatile, and more convenient. As a result, most laboratories have abandoned the use of PC with its large cumbersome glass chambers. Those who have not, continue to use PC because they feel that the original PC methods are superior or because of the lower cost of PC.

The data obtained by planar techniques are reported as R_f values, which were defined and discussed in Chapter 5. Extensive bibliographies have been published.[1,2] Stahl, who first used the name TLC, has written an interesting account of his 20 years in the field[3] as well as a useful laboratory handbook.[4]

The chemicals studied by PC and TLC are the same ones studied by the column LC methods. Hence, it is natural to compare separations performed with these two different formats. For example, a recent paper reports that PC is as good as column LC for three out of five wood sugars, but that LC in columns is better for the other two.[5] Treiber[6] has discussed the utility of TLC and compared it with column LC.

THEORY

The theory of chromatography has been treated in earlier chapters, but the planar methods have some special characteristics that need further discussion.

The Nature of Mobile Phase Flow

In plane chromatography the flow of the mobile liquid cannot be controlled as it can in the column methods. It is dependent on the surface tension γ and the viscosity η of the mobile phase as well as on the nature of the stationary bed. One model considers the bed to be composed of interconnected capillaries of varying diameter. Initially the bed is dry; the liquid is applied at one end and is drawn up the bed against the pull of gravity by capillary action. As a result, the solvent front moves faster than the bulk mobile phase.

A homogeneous bed is preferred to minimize the variation in capillary flow across it. Nevertheless, one would expect that the quantity of liquid on the bed would vary, decreasing from the reservoir to the solvent front. Figure 10.1 shows this phenomenon, and it also shows that the gradient is more pronounced in PC than in TLC, reflecting the greater heterogeneity of paper compared to a thin layer of silica.

The distance the solvent front moves, s_f, has been shown to be proportional to the square root of the migration time t:

$$s_f = \sqrt{kt} \tag{1}$$

where the proportionality constant k is directly proportional to the surface tension and inversely proportional to the viscosity:

$$k = \frac{2K_o d_p \gamma}{\eta \cos \theta} \tag{2}$$

where d_p is the particle diameter and θ is the contact angle (nearly 0 for TLC, so that $\cos \theta$ is unity).

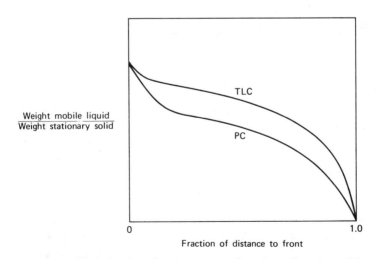

Figure 10.1. Variation in solvent concentration along direction of flow.

The velocity of the solvent front, u_f, is

$$u_f = \frac{k}{2s_f} \tag{3}$$

and is proportional to the surface tension of the mobile phase and inversely proportional to its viscosity and the distance the front has moved. Thus, as noted earlier, the solvent velocity is not constant and decreases the further the solvent has migrated. Eventually, of course, the velocity goes to zero (for ascending methods), which puts an upper limit on the distance the solvent can migrate and determines the maximum size for TLC plates and paper strips. Not only is the flow not constant, but it cannot be easily controlled; the rate is determined by the solvent and the nature of the bed, and the result may not be optimum for good chromatography.

Since the mobile phase is moving on a dry bed, several other undesirable effects occur. The adsorption of the first liquid (at the front) on the stationary phase is exothermic, causing the front to have a higher temperature than the rest of the system. Since the temperature of the system is not usually controlled but is allowed to assume the ambient value, some evaporation may occur at the solvent front. If the solvent is composed of a mixture of liquids, preferential evaporation of the most volatile one will cause a slight change in the solvent composition. In fact, the adsorption of a mixed mobile phase will probably also cause some changes in composition because the most polar component will be preferentially sorbed. The situation can become so severe that solvent demixing can occur. At best, a mixed solvent mobile phase is probably not uniform across the planar bed, and some temperature differentials probably exist as well.

Other Comparisons with Column Methods

All of these uncontrollable factors make it difficult to operate the planar techniques by the best theoretical principles, and their development has been more empirical than the column methods. However, the rate equation predicts that small particle sizes should increase efficiency, and that has been found to be true for TLC. Until a few years ago, the particle size of the silica gel in common TC plates was typically greater than 10 μm in diameter, yielding a few hundred plates for a typical run.

Newer plates have smaller particles of about the same size as used in the high efficiency LC columns—5 to 6 μm, and the plate numbers have risen to more than a thousand. These higher efficiency plates are called *high performance* plates, paralleling the situation in columns, and the technique is known as *HPTLC*. (Like HPLC, the acronymn HPTLC is not recommended.) These plates are thinner than regular plates, the development proceeds to only about half the distance on regular plates, and the separations are faster.[7] They require smaller samples, and streaking has been found to be the best method of sample application.

The silica gel used in TLC has the same properties as that used in LSC in columns, and the discussion about silica in Chapter 9 is relevant. In brief, the silica has a heterogeneous energy surface and many very active silanol groups. It picks up water from the atmosphere very readily and will preferentially adsorb the most polar component in a mobile phase mixture, as just described. Most of the discussion about LSC is also applicable to TLC.

Chromatographic paper contains enough water in its natural form to classify PC as an LLC technique, where water is the stationary phase. Unlike the LLC discussion, however, the mobile phases used in PC are not necessarily immiscible with aqueous solutions.

Thoma and Perisho[8] have summarized the theory for planar chromatography and suggested guidelines for the selection of the best chromatographic conditions. Their main conclusion is that optimum resolution is obtained when R_f is about 0.25 and that a solvent should be chosen to achieve this performance. It is interesting to note that an R_f of 0.25 corresponds to a partition ratio k of 3.0, which is in the optimal range found for column work. A more recent summary of the optimization process has been written by Guiochon.[9]

There is a parameter used in the planar techniques that has not been defined. It was suggested by Martin in 1949 as a way to relate an analyte's structure and its free energy for chromatographic partitioning. It is called R_M and its definition is

$$R_M = \log k = \log \left[\left(\frac{1}{R_f} \right) - 1 \right] \tag{4}$$

By comparing R_M values of compounds that differ by only a methylene group or a functional group, a table of R_M values can be compiled. Since they are proportional to the free energies for these groups, they should be additive, and the total R_M value for any analyte can be predicted by summing the values for the groups of which the molecule is composed. This idea is known as the Martin equation:

$$(R_M)_{tot} = \sum (R_M)_i \tag{5}$$

Its application to PC has been summarized,[10] but this concept does not always work in TLC.[11] Snyder's discussion about R_f values in TLC should be consulted for further details.[11]

THE STATIONARY PHASE

Stationary phases are divided into the same two groups as the techniques in this chapter: paper and thin layers. The paper can be used unsupported, but the thin layers need to be coated on carriers such as glass, aluminum, or plastic. A variety of shapes can be used for the bed in addition to the conventional rectangular shape. Some are circles, wedges, cylinders, rods, and even drums. Thin layers are more flexible and more popular, so they are treated in more detail.

Thin Layers

The most popular thin layer is silica gel, and it is estimated that about 90% of TLC separations are performed on standard TLC silica plates; a common size is 20 cm square and 250 μm thick. Binders, in amounts up to 15%, are usually used to produce a stable layer and good adherence to the backing plate, but plates can be made without binders. Calcium sulfate (gypsum, designated with a G in the name, e.g., silica gel G) and polyvinyl alcohol (PVA) are most common. The PVA plates are very stable and will withstand rather rough handling. While it is not difficult to make one's own plates, most laboratories prefer commercially prepared plates, especially those made with PVA. It must not be forgotten, however, that both binders will modify the adsorption properties of the silica and can produce somewhat different separations.

To facilitate visualization of samples, a phosphor can be added to the plates during preparation. When viewed under UV light, most analytes appear as dark spots against a phosphorescent background and can be located. Here, too, the phosphor becomes a part of the stationary phase, but the inorganic sulfides commonly used are present in small amounts and usually do not alter the separations.

In use, TLC plates are often activated by heating above 100°C for an hour or more. The nature of this activation process and the extent of water removal was presented in Chapter 9. Plates containing organic binder (PVA) should not be heated above 150°C. Dried plates are stored in desiccators to keep them dry and clean. Unfortunately, the plates are exposed to the atmosphere during the time required to apply the sample and get them into the system. Depending on the particular plate, large amounts of water from the atmosphere can be adsorbed on the plate during this time, changing its extent of activation; some plates can absorb more than half their equilibrium concentration of water from a room atmosphere of 50% relative humidity within three minutes. As a result, many chromatographers use commercial TLC plates directly from the box.

Figure 10.2 shows data for three commercial silicas; the type numbers refer to the pore diameters, which are inversely related to the surface areas (Table 1). Type 40 shows a nearly linear behavior while Type 100 shows a nearly constant water uptake between 20 and 70% relative humidity. The specification of the activity of TLC plates can be given according to a system devised by Brockmann[12] whose scale of activities ranges from I for most active to V for least active. Table 2 gives the Brockmann activities for these three silicas at various relative humidities. The laboratory environment can play a significant role in determining plate activity, and, hence, sample separability. If it becomes necessary to determine the Brockman activity of a given adsorbent, the procedure is readily available.[13]

Alumina is the second most popular stationary phase, although the nonpolar bonded phases developed for column use are now available and increasingly used. The operation of TLC in a reverse phase mode is not

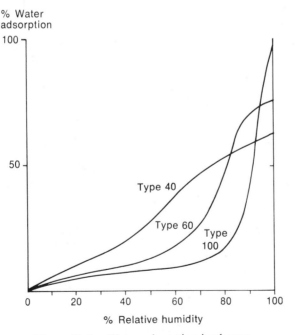

Figure 10.2. Water adsorption isotherms.

TABLE 1   Specifications of the Silica Gels

	Type 40	Type 60	Type 100
Mean pore diameter (Å)	40	60	100
Pore volume (mL/g)	0.65	0.75	1.00
Specific surface (m²/g)	650	500	400

new, but the bonded materials make it more practical and extend the possible applications, including ion-pair chromatography. A recent publication on the use of reverse phase TLC notes that the different stationary phases have different selectivities, but that the major variable in designing a separation is the choice of a mobile phase.[14]

Some difficulty can be experienced with bonded layers containing a large percentage of organic derivative because of its hydrophobic nature. This phenomenon was discussed earlier in the section on bonded phases, but in TLC it may result in the inability of an aqueous mobile phase to move up the plate. If the mobile phase does not wet the plate, the capillary action will be absent. For example, some C_{18} bonded plates are limited to mobile phases containing less than 25% water.

Silica gel can also be used as a solid support on which liquids are immobilized to produce LLC systems. Most common are nonpolar phases for reverse phase work. Other chemicals used as stationary phases include cellulose and cellulose ion exchangers, polyamide, magnesium oxide, and Kieselguhr. Silver nitrate is added to silica gel to retain olefins selectively, as noted earlier. The review by Scott[15] can be consulted for further information on the stationary phase in TLC.

We have already noted the availability of high performance plates for TLC. The other extreme is small plates made on microscope slides for use in fast screening.[15] These slides are easily prepared from a slurry of silica gel in a mixture of methanol and methylene chloride. Directions are available in reference 11 as well as Peifer's original paper.[16] A fourth type

TABLE 2   Brockmann Activity versus Relative Humidity

Relative Humidity	Type 40	Type 60	Type 100
0%	I	I–II	II
20%	II	II	II–III
40%	III	III	III
60%	IV–V	III–IV	III–IV
80%	V	V	IV

of TLC plate has a thick layer for preparative work, and all four are compared in Table 3.

Several special layers are commercially available. One end of the plate can contain a *preadsorption layer*, which is a zone of solid that will not retain the sample. When the sample is spotted in this zone, it will be carried by the mobile phase to the silica zone so that it arrives there in a minimum volume as a small sample. This procedure provides small sample volumes without requiring careful application of small samples. Figure 10.3 shows the cross section of a preparative plate where this preadsorption layer has been combined with a wedge shape for the active layer. The wedge shape will facilitate good separations of larger samples by allowing the analytes to spread out in the third dimension, thus keeping the zone spreading minimal along the axis of migration.

Another type of plate contains two layers, both of which are active but very different in polarity. Figure 10.4 shows a plate composed of a thin layer of nonpolar bonded phase and a larger area of silica gel. These plates are used in two-dimensional TLC with two different mobile phases. The separating power of both normal and reverse phase LC are combined in one plate. The separation of 13 sulfonamides on such a plate is shown in Figure 10.5.

Paper

Although ordinary filter papers can be used, special papers are manufactured for PC. They are highly purified (especially from metal contaminants) and are manufactured under controlled conditions so that such properties as porosity, thickness, and arrangement of cellulose fibers are reproducible from batch to batch. A single paper, Whatman No. 1 or its equivalent, is satisfactory for most separations. The theoretical aspects related to PC have been summarized by Stewart.[17]

As already noted, undried paper contains enough adsorbed water to be classified as an absorption process. Other liquids can be applied to change its characteristics. For example, silicone oils, petroleum jelly,

TABLE 3 Comparison of Types of TLC Plates

Type	Thickness	Particle Size	Sample Size
Regular	250 μm	10–12 μm	≥1 μL
Microscope slides	≤250 μm	10–12 μm	≤1 μL
High performance	±150 μm	5–6 μm	50–500 nL
Preparative	0.5–2 mm	5–40 μm	±150 μL (streaked)

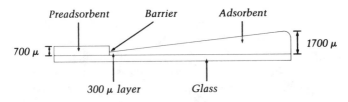

Figure 10.3. Cross-section of tapered prep TLC plate. Courtesy of Analtech. Patent No. 4,348,286, Sept. 7, 1982.

paraffin oil, and rubber latex are used as nonpolar phases. Special papers are also commercially available that contain adsorbents or ion exchange resins or are specially treated (e.g., acylated) or are made of other fibers (e.g., glass, nylon). More details on PC can be found in the monograph by Sherma and Zweig.[18]

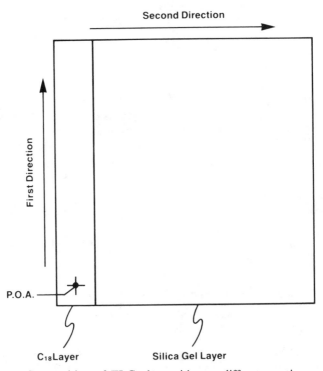

Figure 10.4. Composition of TLC plate with two different stationary phases. Courtesy Whatman, Inc.

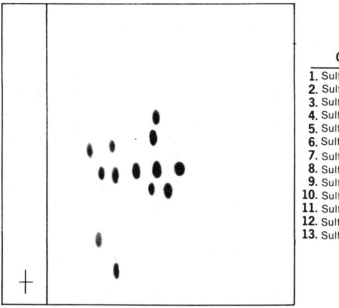

Compounds

1. Sulfisoxazole
2. Sulfathiazole
3. Sulfadiazine
4. Sulfaquinoxaline
5. Sulfachlorpyridazine
6. Sulfaguanadine
7. Sulfamerazine
8. Sulfabromethazine
9. Sulfadimethoxine
10. Sulfamethazine
11. Sulfaethoxypyridazine
12. Sulfanilamide
13. Sulfapyridine

SULFONAMIDES ON WHATMAN MULTI-K PLATE

Plate: Multi-K; C_{18} strip on silica gel plate; 20 x 20 cm

Sample Data —
Preparation: 13 Sulfonamides, 1 mg/ml in
  Acetone/Methanol (90:10)
Applied Volume: Approx. 10 μl applied to C_{18} strip

Development: 2-Dimensional
  (1) C_{18} development to 16 cm in equilibrated tank
  (2) Silica gel development to 14 cm in
    unequilibrated tank

Solvent: (1) (C_{18} layer): Toluene/CH_3CN (80:20)
       (2) (Silica gel layer): Ethyl acetate/MeOH/
         NH_4OH (85:15:0.6)

Fig. 10.5. 2-D separation of 13 sulfonamides on Multi-K plate. Courtesy Whatman, Inc.

THE MOBILE PHASE

Frequently the mobile phase is a mixture of liquids chosen and optimized by trial and error because results can be obtained so quickly and easily. The principles of mobile phase selection discussed in the last chapter are, of course, relevant. For normal phase systems on silica gel, a nonpolar liquid is modified with a more polar one, and small amounts of a third component like acetic acid are often added to deactivate the plate slightly and decrease tailing. The amount of water is critical in determining the activity of the plate, as already indicated. Many times the chosen mobile phase is simply a mixture that works, and a typical example is a mixture of butanol/acetic acid/water. For reverse phase systems, a polar mixture should be used.

Because these solvents are used in the open, evaporation can be a problem. Continued use of a mobile phase solvent mixture will probably result in preferential evaporation of the most volatile component, thus changing the composition. Demixing can also occur on the bed if the solvents do not have a high degree of mutual solubility. Gradient techniques can be used in the mobile phase and in the stationary phase.[19]

APPARATUS AND PROCEDURES

TLC and PC can be run very simply. The steps are: sample application, development (the name used for the actual running of the chromatogram), visualization and detection, and quantitation. As each of these is discussed, the apparatus required will also be included.

Sampling

As in all chromatographic processes, the sample should occupy as small a volume on the bed as possible. Solutions of the sample can be applied as spots or streaks at one end of the bed. In neither case should the application of sample disturb the bed, so that for regular TLC plates the sampling device cannot touch the surface. The harder layers on some commercial plates are preferred for this reason.

Spotting is done with micropipets or simple capillary glass tubing. A template facilitates the process. As noted in Table 3, the sizes can be quite small and contain about 50 to 100 µg of sample. To keep the spots small, sample can be applied repetitively to the same area, allowing the previous sample to dry before reapplication. Plates with the preadsorption layer are easy to use, and they deposit the sample as a very narrow zone at the edge of the active layer. In general, the placement of the spots must be far enough from the end of the bed to prevent it from dipping into the solvent reservoir (1 to 2 cm). The actual distance can have a significant effect on R_f values, especially in PC with mixed solvents.[20]

Automated sample spotters are commercially available that can apply exact volumes to multiple spots simultaneously. In his recent review of instrumentation, Beesley[21] discussed four of them.

Streaking is used primarily in prep TLC, and automatic devices are also available for this purpose.[21] This process has the same objectives and requirements as spotting, but it permits larger amounts to be applied and uses the entire bed (no empty spaces between spots).

As robotics come into greater use, these automatic devices will undoubtedly become incorporated into a fully automated TLC process.

Development

The prepared plate or paper is developed in a closed, presaturated chamber using an ascending or descending mobile phase flow. Four common configurations are shown in Figure 10.6. Paper requires support, as shown

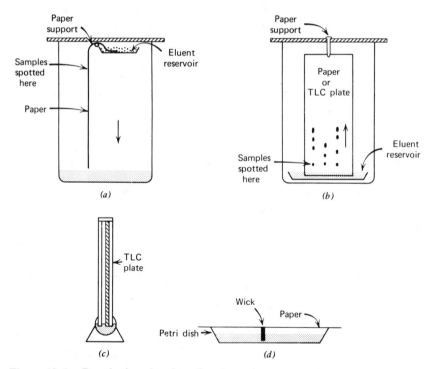

Figure 10.6. Developing chambers for plane chromatography: (a) descending—used with PC; (b) ascending—used with TLC and PC; (c) sandwich—used with TLC; (d) horizontal—used with paper as shown, but also adaptable for HPTLC.

in (a), (b), and (d) in the figure. For the large chambers, (a) and (b), large filter papers are inserted and wet with solvent in order to presaturate the chamber, and all chambers are covered, as shown. Development time varies from about 15 min for TLC to several hours for descending PC.

Special developing chambers are available that allow five different solvents to be run simultaneously; elaborate devices have been marketed for use with the new high efficiency plates. Other operating procedures that can be run on conventional chambers are:

1. Multiple development and programmed multiple development[22,23] in which the plate is developed several times in the same direction with the same or different solvents and dried between developments. Spot reconcentration occurs with each development, keeping it small and improving separations. Reference 22 elaborates on the many variations of this procedure.

2. Overrun, whereby the mobile phase is allowed to run off the end of the bed, thus increasing R_f values that would otherwise be small.

3. Two-dimensional development, in which a second mobile phase is run perpendicular to the first mobile phase flow on a square plate that is dried after the first development. The field has recently been reviewed by Guiochon and co-workers.[24] A variation of this technique using a bed with two stationary phases was shown in Figure 10.4.

Beesley[21] has described some special development chambers designed for nonconventional TLC operation. Camag has produced an automated radial system whose operation closely approximates that used in column LC. A syringe pumps mobile phase into the center of a plate, and a sample is injected into the flowing stream and is separated as concentric circles on the plate.

Newman-Howells Associates manufactures a sandwich chamber that is "overpressurized" and thus allows the control of flow. And, finally, Harrison Research makes a preparative device that uses centrifugal force to move the mobile phase through a radial plate.

Visualization and Detection

After the development is completed and the plate (or paper) dried, the analyte bands must be located unless they themselves are colored. Bobbitt[25] lists 6 universal spray reagents, including sulfuric acid and iodine, and 89 other specific reagents for TLC, and Touchstone and Dobbins[26] describe the use of a total of 207 reagents. If the analytes fluoresce, they can be located under a UV lamp. If not, the use of a phosphor in the TLC plate may make it possible to locate them as nonfluorescing spots on the phosphorescing plate.

For qualitative identification, the R_f values are calculated from the distances measured on the plate or paper. Alternatively, the distances migrated can be measured relative to a standard run on the same plate, which usually is more reproducible. The colors caused by the visualizing reagents can also be used for identification sometimes.

Quantitative Analysis

The spots can be quantitated by removing them from the plate or paper, or they can be measured *in situ*. Three methods of quantitation on the plate are (1) measuring spot size, (2) densitometering, and (3) radioactive counting. The square root of the spot area is approximately proportional to the log of the weight of the analyte. This is a semiquantitative method, but it is fast. Standards are run along with the unknown so that an estimate of the quantity of analyte can be made visually from the plate with about 25% accuracy.

For greater accuracy, the plate is scanned with a densitometer in one of the three modes—transmission, reflectance, or fluorescence. Some instruments can also provide spot areas. Four scanners are described in Beesley's review,[21] and Touchstone and Sherma[27] have described the use of densitometers. Accuracies are about 2 to 5%.

If the sample is radioactively tagged, it can be quantitated with one of the radioscanning devices. Reference 21 provides more details of all the scanners, including photographs of them.

Spots can be scraped off TLC plates or cut out of paper chromatograms and extracted to remove the analytes. Then they can be determined by any conventional technique; UV/visible spectrophotometry is popular.

EVALUATION

TLC is the more popular of the two planar techniques. It is used for fast screening of mixtures and for surveying methods for use in columns. Quantitative analysis is better accomplished by column LC, but quantitative TLC is increasing in popularity as the instrumentation improves. Table 4 contains a summary of the advantages and disadvantages of the planar methods.

The biennial review in *Analytical Chemistry*[28] is a good source of information on the latest developments in planar chromatography.

TABLE 4   Advantages and Disadvantages of Planar Chromatography

Advantages	Disadvantages
Fast	Flow is not constant
Simple, inexpensive apparatus	Flow cannot be controlled easily
High sample throughput	Limited plate numbers
Flexible shapes	Not always predictable or in
Two-dimensional	agreement with theory
Detection and Quantitation is static	Temperature gradients can exist
All sample components are in	Quantitation not as accurate as for
evidence	column LC

REFERENCES

1. K. Macek, I. M. Hais, J. Kopecky, and J. Gasparic, *Bibliography of Paper Chromatography and Thin Layer Chromatography, 1961–65*, supplementary volume, *J. Chromatogr.*, Elsevier, Amsterdam, 1968.

2. D. Janchen (ed.), *Thin Layer Chromatography: A Cumulative Bibliography in Three Parts, 1965–1973*, Camag, Muttenz, Switzerland, 1974.

3. E. Stahl, *J. Chromatogr.* **1979**, *165*, 59.

4. E. Stahl, *Thin Layer Chromatography*, 2d ed., Springer, New York, 1969.

5. R. C. Peltersen, V. H. Schwuadt, and M. J. Effland, *J. Chromatogr. Sci.* **1984**, *22*, 478.

6. L. R. Treiber, *J. Chromatogr. Sci.* **1986**, *24*, 220.

7. C. F. Poole, S. Khatib, and T. A. Dean, *Chromatogr. Forum* **1986**, *1*(1), 27.

8. J. A. Thoma and C. R. Perisho, *Anal. Chem.* **1967**, *39*, 745.

9. G. Guiochon et al., *J. Chromatogr. Sci.* **1978**, *16*, 152, 470, 598; **1979**, *17*, 368.

10. J. Green and D. McHale, *Adv. Chromatogr. N. Y.* **1966**, *2*, 99.

11. L. R. Snyder, *Adv. Chromatogr. N. Y.* **1967**, *4*, 3.

12. H. Brockmann, *Angew. Chem.* **1947**, *59*, 199.

13. R. J. Gritter, J. M. Bobbitt, and A. E. Schwarting, *Introduction to Chromatography*, 2d ed., Holden-Day, Oakland, Calif. 1985.

14. E. Heilweil and F. M. Rabel, *J. Chromatogr. Sci.* **1985**, *23*, 101.

15. R. M. Scott, *J. Chromatogr. Sci.* **1973**, *11*, 129.

16. J. J. Peifer, *Mikrochim. Acta* **1962**, 529.

17. G. H. Stewart, *Adv. Chromatogr. N. Y.* **1965**, *1*, 93.

18. J. Sherma and G. Zweig, *Paper Chromatography and Electrophoresis*, Vol. 2, *Paper Chromatography*, Academic Press, New York, 1971.

19. A. Niederwieser and C. C. Honegger, *Adv. Chromatogr. N. Y.* **1966**, *2*, 123.

20. H. G. Cassidy, *Fundamentals of Chromatography*, Vol. 10 of *Technique of Organic Chemistry*, A. Weissberger (ed.), Wiley-Interscience, New York, 1957, p. 161.

21. T. E. Beesley, *J. Chromatogr. Sci.* **1985,** *23,* 525.

22. T. H. Jupille and J. A. Perry, *Science* **1976,** *194,* 288.

23. J. A. Perry, *J. Chromatogr.* **1975,** *113,* 267.

24. M. Zakaria, M.-F. Gonnord and G. Guiochon, *J. Chromatogr.* **1983,** *271,* 127.

25. J. M. Bobbitt, *Thin-Layer Chromatography*, Reinhold, New York, 1963.

26. J. C. Touchstone and M. F. Dobbins, *Practice of Thin Layer Chromatography,* 2d ed., Wiley-Interscience, New York, 1983.

27. J. C. Touchstone and J. Sherma, *Densitometry in Thin Layer Chromatography: Practice and Applications,* Wiley, New York, 1979.

28. J. Sherma, *Anal. Chem.* **1986,** *58,* 69R.

SELECTED BIBLIOGRAPHY

Kirchner, J. G., in *Techniques of Chemistry*, Vol. 14, *Thin-Layer Chromatography,* 2d ed., E. S. Perry (ed.), Wiley-Interscience, New York, 1978.

Shellard, E. J. (ed.), *Quantitative Paper and Thin Layer Chromatography,* Academic Press, New York, 1969.

Touchstone, J. C. (ed.), *Advances in Thin Layer Chromatography: Clinical and Environmental Applications,* Wiley, New York, 1982; *Practice of Thin Layer Chromatography,* 2d ed., Wiley, New York, 1983; *Quantitative Thin Layer Chromatography,* Wiley, New York, 1973.

Touchstone, J. C. and J. Sherma (eds.), *Techniques and Applications of Thin Layer Chromatography,* Wiley, New York, 1985.

Zlatkis, A. and R. E. Kaiser, *HPTLC-High Performance Thin Layer Chromatography,* Elsevier, New York, 1977.

OTHER TOPICS

There are a few important topics that do not fit in the other chapters but do not merit complete chapters of their own. Other topics could fit in several chapters, but they have been deferred until this chapter so that a complete discussion can be given in only one place. Each of them will be treated briefly.

SUPERCRITICAL FLUID CHROMATOGRAPHY (SFC)

This topic was introduced in Chapter 5, where a supercritical fluid was defined and shown on a phase diagram. A supercritical fluid is neither a gas nor a liquid, but a truly unique phase with properties intermediate between the two. As one would expect, then, SFC is a hybrid of GC and LC.

The development of SFC up to 1983 has been briefly described by Gere,[1] showing the evolution in concept and the early applications. The first commercial instrument was offered for sale in that year, giving considerable impetus to the field. Although that instrument was discontinued, several new ones have replaced it, and the number of publications in SFC is rising quickly. In 1986, the *Journal of Chromatographic Science* devoted the entire June issue to SFC.[2] The newest instruments are designed to use OT columns similar to those used in GC,[3] and, while the relative advantages of packed versus OT columns are still being investigated, the OT type have stimulated a great deal of interest and activity.[4] Most chromatographers expect a large increase in interest in SFC in the coming years. Whether or not the field has yet become "routine," one paper has been written with that title,[5] and the status of available instruments has been reported in two papers,[3,6] the latter including a review of the papers at the 1986 Pittsburgh conference.

Supercritical fluids are used for other analytical purposes, including extraction. Gere[1] and Berry[6] have also discussed them in their papers.

Principles of Operation

The properties of supercritical fluids fall between those of gases and liquids, as shown in Table 1. Thus the mobile phase in SFC has a viscosity

TABLE 1 Typical Values of Parameters Important in Chromatographic Zone Broadening

Parameter	Approximate Value		
	GC	SFC	LC
Diffusion coefficient (cm^2/sec)	10^{-1}	10^{-4}	10^{-5}
Density (g/mL)	10^{-3}	0.8	1
Viscosity (g/cm-sec)	10^{-4}	5×10^{-4}	10^{-2}

only slightly greater than the gases in GC while showing much more affinity for analytes because of its higher density. On the other hand, a supercritical fluid has a higher diffusion coefficient than a liquid, making SFC more efficient than LC. In some respects, SFC has the best characteristics of GC and LC, the two extremes.

Mobile Phase Properties. Some chemicals that could be used in SFC are listed in Table 2. The one that has been used most commonly is carbon dioxide, and it will be the focus of this short introduction. Figure 11.1 shows the pressure–volume phase diagram for CO_2 at various temperatures. The critical values (P_c = 7.4 MPa, V_c = 96 mL, and T_c = 31° C) intersect approximately at the point marked X. Liquid exists in the lined space at the left of the diagram, gas and liquid are in equilibrium in the space cut off by the dashed line, supercritical fluid exists above the critical temperature, and gas exists at the right. Remember that the critical temperature is that temperature above which a gas cannot be liquefied no matter how high the pressure.

If a chemical such as CO_2 is above its critical temperature and the pressure is raised, the density of the fluid increases, as shown in Figure 11.2. The operating region for SFC is in the center of the figure, above the gas–liquid equilibrium region. Most systems are run at constant temperature; if the pressure is increased during a run, the density will increase according to the isotherms shown. Computer programs can linearize density changes to achieve linear density programming.

The viscosity of a supercritical fluid is constant at constant density, regardless of the temperature and pressure. It increases with density, but it is still lower than liquid viscosities by a factor of ten or more. As a consequence, pressure drops across SFC columns are smaller than across LC columns, requiring less pressure for a given flow and making high velocities realistic. OT columns can be operated at pressures only slightly above the critical value of 7 MPa (73 atm or 1100 psi), and even packed columns show pressure drops as low as 1.5 MPa, resulting in inlet pres-

TABLE 2   Properties of Possible Mobile Phases for SFC[a]

Compound	Atm b.p. (°C)	Critical Point Data		
		T_c (°C)	P_c (MPa)	d_c (g/mL)
Nitrous oxide	−89	36.5	7.23	0.457
Carbon dioxide	−78.5[b]	31.3	7.38	0.448
Sulfur dioxide	−10	157.5	7.86	0.524
Sulfur hexafluoride	−63.8[b]	45.6	3.76	0.752
Ammonia	−33.4	132.3	11.3	0.24
Water	100	374.4	23.0	0.344
Methanol	64.7	240.5	7.99	0.272
Ethanol	78.4	243.4	6.38	0.276
Isopropanol	82.5	235.3	4.76	0.273
Ethane	−88	32.4	4.89	0.203
n-Propane	−44.5	96.8	4.25	0.220
n-Butane	−0.5	152.0	3.80	0.228
n-Pentane	36.3	196.6	3.37	0.232
n-Hexane	69.0	234.2	3.00	0.234
n-Heptane	98.4	267.0	2.74	0.235
2,3-Dimethylbutane	58.0	226.8	3.14	0.241
Benzene	80.1	288.9	4.89	0.302
Diethyl ether	34.6	193.6	3.68	0.267
Methyl ethyl ether	7.6	164.7	4.40	0.272
Dichlorodifluoromethane	−29.8	111.7	3.99	0.558
Dichlorofluoromethane	8.9	178.5	5.17	0.522
Trichlorofluoromethane	23.7	196.6	4.22	0.554
Dichlorotetrafluoroethane	3.5	146.1	3.60	0.582

[a] From Gouw and Jentoft, Adv. Chromatogr. N.Y. 1975, 13, 1.
[b] Sublimation point.

sures around 8.5 MPa (1300 psi). These values are readily achieved with available pumps and are not the dangerously high values originally anticipated for SFC.

Carbon dioxide is not very polar, and this characteristic has been a limitation in running polar analytes. Table 2 shows that there are not very many polar chemicals that could be considered for use in the supercritical region. A popular alternative is to add methanol as a modifier to the carbon dioxide. Figure 11.3 shows the effect of adding 1% methanol to the CO_2 in the separation of glycerides; peak shapes are improved and the retention times are decreased.

Stationary Phase Properties. Column stability requires that the stationary phases be bonded to the column wall (OT) or the solid support (packed).

The packings originally made for LC are used, and special bonded and crosslinked OT columns are becoming available.

Some problems have been experienced due to swelling of the stationary phase by the supercritical fluid. This is especially serious for columns that have high stationary loads, and it affects the column performance.[7] More experience is needed, and new columns may be required.

To date, much work has been done with the nonpolar carbon dioxide on nonpolar stationary phases. Better systems will certainly be found. Some theoretical work has been done and can be found in the references cited. Peaden and Lee[8] have discussed resolution in OT columns, and SFC is included in *Analytical Chemistry*'s biennial review on GC.[9]

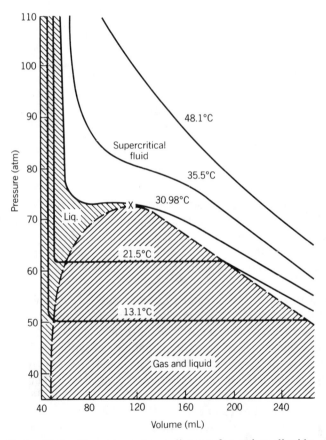

Figure 11.1.   Pressure–volume diagram for carbon dioxide.

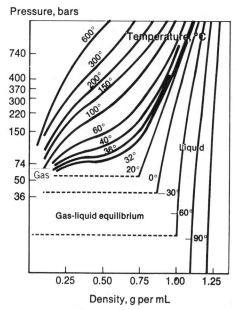

Figure 11.2. Pressure–density diagram for carbon dioxide. Reprinted with permission from *Chem. Eng. News* **1982**, *60*(12), 46. Copyright 1982, American Chemical Society.

Instrumentation

SFC instruments are hybrids of LC and GC instruments, and early commercial models were modifications of one or the other. If packed columns are used, the instrument is more like a liquid chromatograph with a reciprocating pump and a pressurized UV detector. If OT columns are used, the instrument is more like a gas chromatograph with a syringe pump and an FID.

In both types, a pressure restriction is needed somewhere after the column in order to keep the pressure above the critical value. In some cases it comes after the detector, which is then operated much as it is in LC, except that it has to have the capability of withstanding the higher pressure. If it comes before the detector, some problems have arisen as the depressurized effluent enters the detector. Noise is generated and is believed to originate with the formation of small nonvolatile particles that can also plug the transfer lines.

Two methods have been reported for using a gas chromatographic FID. In one case, the flame jet is crimped to maintain the necessary pressure up to that point.[10] In the other case, a 9 cm × 10 μm capillary is used to connect the column with the FID,[11] and it serves as the necessary restriction at the end of the column without resulting in the noise mentioned above. The performance of capillary restrictors has been thoroughly explored,[12] and temperatures at least 1.3 times the critical temperature are recommended to avoid these solvent clusters and demixing.

The OT columns presently in use have smaller internal diameters than those normally used in GC—typical values are 50 and 100 μm. Lengths can be longer than in GC, but usually they are about the same, and efficiencies are similar. The temperature must be maintained above the

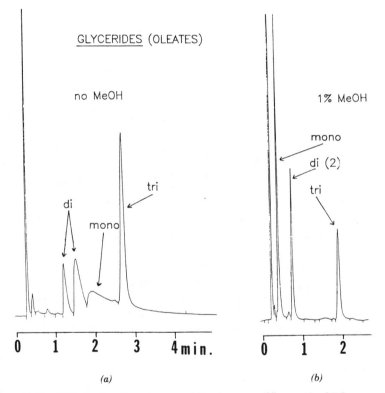

(a)   (b)

Figure 11.3.   Effect of methanol as a mobile phase modifier on the SFC separation of mono-, di-, and triglycerides of oleic acid. Reprinted from *American Laboratory*, **1984**, *16*(5), 19. Copyright 1984 by International Scientific Communications, Inc.

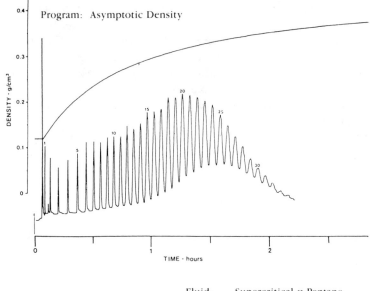

Fluid: Supercritical *n*-Pentane
at 210 °C
Column: 10m x 100μm, i.d. SB-
Phenyl-50
Detector: UV Absorbance

Figure 11.4.   Example of asymptotic density programming in SFC. Polystyrene oligomer mixture with avg. MW of 2,000. Reproduced from the *Journal of Chromatographic Science* by permission of Preston Publications, Inc.

critical value, and most applications (with CO_2) are around 50 to 150°C. Flow rates are on the range of several μL per minute.

Analyte solubility is directly related to density, so density programming is similar in its effect to programmed temperature in GC. Density programming is facilitated if the column has a low pressure drop, which is one reason for the popularity of OT columns. Pulseless pumps are also desirable. Asymptotic density programming, rather than linear programming, produces equal spacings between homologs, as shown in Figure 11.4.

Applications

SFC is best suited to separate materials that are not volatile enough for GC and are not easily handled by LC because of problems like suitable detectors. Thus high molecular weight materials that do not absorb UV and cannot be used with UV detectors in LC are commonly run by SFC. One example is shown in Figure 11.5; the oligomers in a polymeric detergent mixture are separated using pressure programming.

Coal tar is a mixture that needs the high efficiency of OT columns and approaches the upper temperature limit of GC. Figure 11.6 compares a separation of coal tar components by GC and SFC. As expected, the GC is performed with programmed temperature and the SFC with programmed density, as indicated in the figure. These are only a few examples, and many more are expected in the next few years.

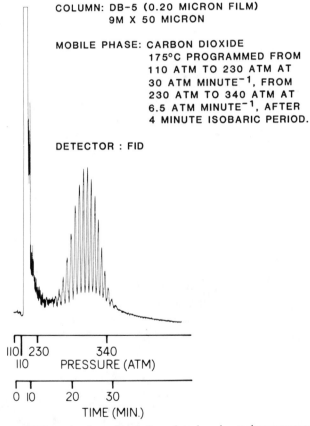

COLUMN: DB-5 (0.20 MICRON FILM)
9M X 50 MICRON

MOBILE PHASE: CARBON DIOXIDE
175°C PROGRAMMED FROM
110 ATM TO 230 ATM AT
30 ATM MINUTE^{-1}, FROM
230 ATM TO 340 ATM AT
6.5 ATM MINUTE^{-1}, AFTER
4 MINUTE ISOBARIC PERIOD.

DETECTOR : FID

110 230   340
110   PRESSURE (ATM)

0 10   20   30
        TIME (MIN.)

Figure 11.5. SFC separation of an ethoxylated amine using pressure programming. Courtesy of Suprex.

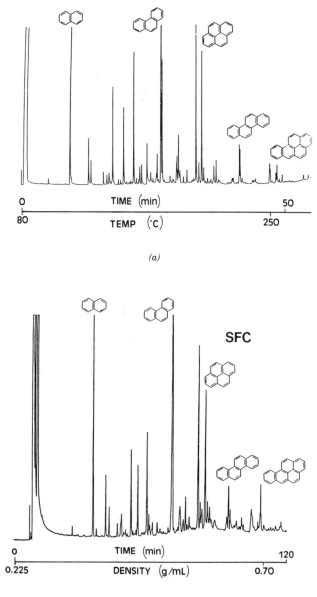

Figure 11.6. Comparison of a coal tar separation by (*a*) GC and (*b*) SFC. Both separations performed on SE-54 OT columns with 0.25 μm film thicknesses. GC column 20 m × 300 μm, temperature programmed from 40°C to 265°C. SFC column 34 m × 50 μm, density programmed. Reprinted with permission from J. C. Feldsted and M. L. Lee, *Anal. Chem.* **1984,** *56,* 619A. Copyright 1984, American Chemical Society.

CHROMATOGRAPHY AND ON-LINE MASS SPEC AND INFRARED

In Chapter 6 we saw that, by itself, chromatography is not well suited to qualitative analysis; thus it is often combined with other methods. The most successful combination has been GC with mass spectrometry (MS); GC's ability to separate materials and MS's ability to identify them has made the combination one of the most powerful analytical techniques available today. The other forms of chromatography are also being combined with MS and with infrared spectroscopy (IR). The resulting analytical methods are usually designated by their combined abbreviations (e.g., GC/MS or GC-MS) and are known as *hyphenated techniques*. The current status of these methods will be described briefly.

All three chromatographic techniques—GC, LC, and SFC—have been combined with MS and with IR. Not all have been equally developed, and none has yet reached the reliability of GC/MS. Additional combinations are possible, and chromatography will probably be combined with other measuring instruments in the future.

Requirements for Interfacing

The development of hyphenated techniques has depended on advances in at least three areas: interfacing, scanning speed of the measuring technique, and adequate data systems. The different combinations of chromatography and on-line measuring instruments have different requirements, but a few generalizations can be made. Since the GC/MS combination is the most popular and successful, it will be described as a typical example of the types of considerations required for interfacing instruments. Problems associated with other interfaces will be mentioned briefly later.

The effluent from the chromatograph may not be directly compatible with the inlet for the measuring instrument. For example, the mass spectrometer operates under high vacuum (0.01 Pa) and cannot be connected directly to a chromatograph whose column effluent is at atmospheric pressure (100 kPa). It is necessary to reduce this pressure to that of the mass spectrometer in order to couple them together. Furthermore, it would be desirable if this reduction in pressure could be accomplished without reducing the analyte concentration. The problem of getting rid of the mobile phase is even more drastic in LC/MS. Other interfacing problems have to do with relative volatilities and the states of matter most easily handled by the various techniques.

The second requirement is fast scanning so that the spectra of analytes can be obtained "on the fly." For fast peaks like those obtained with OT columns, scans must be completed on the order of milliseconds. Quadrapole mass specs and Fourier transform infrareds (FTIR) have developed into reliable fast-scanning instruments suitable for interfacing with chromatographs.

A final requirement is a data system capable of handling all the data generated by a fast-scanning MS or FTIR. The development of reliable, small, and inexpensive microcomputers paralleled the development of GC/MS and subsequent instrumental combinations and made them realistic in terms of complexity and cost.

GC/MS and GC/FTIR

Both these techniques, GC/MS and GC/FTIR,[13] are highly developed, and many commercial instruments are available. The special requirements of each will be discussed briefly.

The GC/MS interface requirements have already been mentioned. The early devices attempted to remove the light mobile phase gas and enrich the effluent in analyte concentration. Semipermeable membranes and sintered glass were used, and the most popular devices carried the names of their inventors—Watson–Biemann, Llewellyn–Littlejohn, and Ryhage.[14] Since the discrepancy in pressures is lessened if the GC flow is decreased, the recent popularity of OT columns has opened up new possibilities for interfaces. Interfaces used with OT gas chromatographs can be very simple, and two types are common. One is called *direct*, and that is what it is. The end of the OT column is extended from the GC directly into the ion source of the MS. The GC flows are low enough, and the vacuum pumping high enough that the vacuum required by the mass spectrometer can be maintained without any other interfaces or attachments. In this arrangement, the outlet of the GC column is operated at vacuum. This is not common, but it has been shown that vacuum operation is not deleterious but advantageous.[15] While the direct interface is very convenient, it has two disadvantages: all column effluent is deposited in the ion source of the mass spectrometer, causing it to become contaminated rather quickly, and the GC column cannot be changed without shutting down the MS because there is no way to isolate one from the other.

The other arrangement is called *open split* and is shown in Figure 11.7. The space between the column and the MS inlet is maintained at about atmospheric pressure by the use of a second source of gas and a separate vacuum. By controlling the amount of purge gas in this region, the column can be disconnected without shutting down the MS, and, alternatively, undesirable sample components such as large quantities of solvent can be removed before they enter the MS.

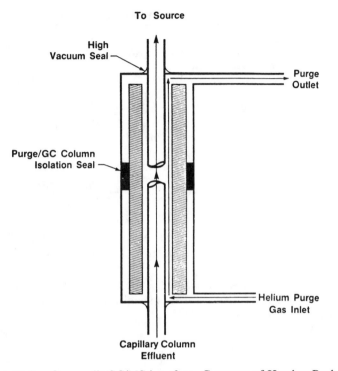

Figure 11.7.   Open split GC/MS interface. Courtesy of Hewlett-Packard.

The situation is different for GC/FTIR. In this case, large samples are desired because they facilitate detection by IR, but the vapor state analytes are not easily sampled by IR. A light pipe has been designed that can be heated to keep the analytes in the vapor state and also has the following requirements of a good IR cell for use with GC: small volume, long path length, and high transmission. A typical light pipe is 50 cm × 1 mm in size and has a reflecting gold coating. Its description and details of its use including the data handling requirements have been discussed.[16]

The other interface is a type of matrix isolation in which the analytes are frozen. A comparison of these two types has recently been published.[17] Both are listed in Table 3 along with the other interfaces.

A good overview of the operation of a GC/MS/DS can be found in the *Journal of Chemical Education*.[18] The recent reviews of MS[19] and IR[20] contain the latest references on GC/MS and GC/IR, respectively.

GC/FTIR has become as sensitive as GC/MS, which has been highly regarded for its excellent sensitivity. The two techniques provide different data and are complementary; IR is particularly advantageous for identi-

TABLE 3   Some Common Interfaces for Hyphenated Techniques

	Interfaces Used with	
Chromatographic Technique	MS	IR
GC	Direct	Light pipe
	Open split	Matrix isolation
LC	Thermospray	ATR
	DLI	KBr pellets
	Moving belt	
SFC	DFI	Cell, high pressure
	Molecular beam	ATR
		KBr pellets

fying isomers that cannot be distinguished by MS. MS offers the possibility of quantitative analysis by the isotope dilution method, which makes it uniquely attractive.

Since IR is nondestructive, it is possible to combine the three instruments into one: GC/FTIR/MS. The special requirements and some applications have been described.[21]

LC/MS and LC/FTIR

For LC/MS the main problem is the large amount of mobile liquid phase that must be removed to get the effluent reduced to the high vacuum of the MS. Microbore columns are desirable for this reason.[22] The three most popular devices have been summarized by Majors[23]: direct liquid interface (DLI), moving belt transport, and thermospray.[24] The thermospray device consists of a small bore capillary tube that is heated to produce a stable, high-velocity jet consisting mostly of vapor with a small amount of mist. It not only provides an interface to the MS, it also causes the ionization of analytes necessary for the MS. Some think it may find more widespread use as a transport device.

The latest references on LC/MS are reviewed in references 19, 22, and 25, and early interfaces are discussed in reference 14.

The liquid from an LC is compatible with normal IR sampling and is less of a problem. However, LC mobile phases may not be transparent in the IR, and water is a particularly difficult solvent to handle. For volatile organic solvents, evaporation is possible, and the remaining nonvolatile analytes can be deposited in KBr and pressed into pellets. Further discussion can be found in reference 20.

SFC/MS and SFC/FTIR

Since SFC is in its infancy, the same is true of the hyphenated techniques that involve it. In general, it would be expected that SCF/MS should use interfaces like GC/MS since the supercritical fluid mobile phases become gases when reduced to atmospheric pressure, but the conditions are more severe because of the higher (critical) pressure. OT columns, because of their low pressure drops, are favored. The two interfaces that have been used are a direct fluid injection (DFI) and a molecular beam apparatus. DFI has been used with packed columns[26] and with OT columns,[27] using both chemical ionization and electron impact ionization. For a more complete discussion of both interfaces, see the chapter on SFC in the ACS Symposium Series edited by Ahuja[28]; the recent review on LC/MS[25] also contains considerable information about SFC/MS.

For SFC/IR it would seem logical to try to keep the mobile phase as a fluid and use interfaces similar to those used in LC/IR. The common SFC mobile phase CO_2 is sufficiently transparent in the IR to serve as a solvent in that fashion. Alternatively the CO_2 can be easily volatilized, providing the opportunity to use interfaces like those in GC/IR. A few references are provided in the IR review,[20] but new devices are likely in the next few years as SFC matures.

CHIRAL SEPARATIONS

It is becoming increasingly important to distinguish between enantiomers for optically active compounds. For many drugs, only one optical isomer is pharmacologically active, and a total analysis that does not separate and quantitate the enantiomers is unsatisfactory. Unfortunately for the chromatographer, enantiomers have identical physical properties, and it is unrealistic to expect to separate them with conventional chromatographic systems.

Traditionally this problem has been solved by various methods, including the formation of diastereomers using a chiral reagent that creates two chiral centers in the products. Diastereomers have different physical properties, and they can be separated with conventional nonchiral chromatographic systems. This approach was the major one used in GC when the field began to develop about 15 years ago. However, it has some disadvantages, and GC is not as suitable as LC for separating nonvolatile drugs, which are of considerable interest. More recently, the major advances have been made in LC using chiral phases.

Four reviews that have appeared in *Advances in Chromatography* between 1974 and 1983 chronicle the developments. The first review in 1974 covered GC separations and emphasized diastereomer separations.[29] The next three reported LC separations, mostly of enantiomers; the 1978 review by Krull[30] covered all types, but the 1980 review by Davankov[31] described only ligand exchange. In 1983, Davankov[32] updated his review of ligand exchange methods and included some examples of other types of LC separations. Most of the work since then has been on the preparation of chiral stationary supports.

Diastereomer Separations

If a racemic mixture of optical isomers (designated *R* and *S*) is reacted with a chiral reagent (designated, e.g., *R'*), the reaction products will have two chiral centers and can be designated as *RR'* and *SR'*. These two diastereomers have different physical properties and can be separated chromatographically with nonchiral phases.

If the derivatization is carried out with a chiral reagent composed of both optical isomers (designated *R'* and *S'*), four reaction products are obtained. Their relationships are shown in Figure 11.8. If a chromatographic separation of these four isomers is attempted with an achiral system, two peaks can usually be obtained (Figure 11.9) representing the diastereometric forms—*RS'* and *SR'* in one peak and *RR'* and *SS'* in the other. If this mixture is run on a chiral chromatographic system, all four compounds may be resolved, as shown in Figure 11.9*b*.

Usually the objective in forming diastereomers is to use a chiral reagent that is 100% of one isomer and to use an achiral chromatographic system. As an example, Clark and Barksdale[33] separated the *R* and *S* enantiomers of amphetamine after reacting the racemic mixture with 1[(4-nitro-

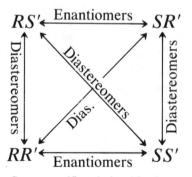

Figure 11.8. Stereospecific relationships between isomers.

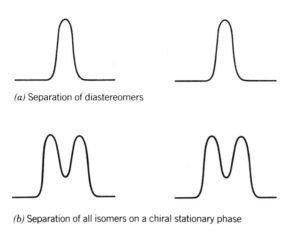

(a) Separation of diastereomers

(b) Separation of all isomers on a chiral stationary phase

Figure 11.9.   Typical separations of stereoisomers.

phenyl)sulfonyl]prolyl chloride, which was 100% of the S isomer. The RS and SS diastereometric amides were separated on a silica gel column by LC using a chloroform/heptane (8/2) mobile phase. The chromatogram is shown in Figure 11.10 along with the structure of the derivatives. In another recent example, the enantiomers of 21 amino acids were separated after they were derivatized with a chiral reagent to form the diastereomers.[34] The review by Lochmuller and Souter[35] summarizes the early GC work on enantiomer separations, including a comprehensive discussion of the use of liquid crystals as stationary phases.

For GC, the two most useful chiral reagents are N-trifluoroacetyl-L-prolyl chloride (TCP) and menthyl chloroformate (MCF). TCP is used for the resolution of amines and MCF for alcohols. At best the reagents are 99% optically pure, but they can be as little as 90% pure.

The necessity to use an optically pure reagent is one disadvantage of this method of separation. Other disadvantages are: the reagent must react equally with the two enantiomers in the sample; the separated diastereomers must be reacted to remove the unwanted derivatizing reagent; and then the separated enantiomers must be purified. For these reasons, this approach has not been the most successful or popular. It is easier to use a chiral chromatographic system.

Chiral Mobile Phases

Either the mobile phase or the stationary phase can be chiral in order to effect the desired separation of enantiomers. The use of chiral mobile phases assumes that the analytes form strong associations with a chiral

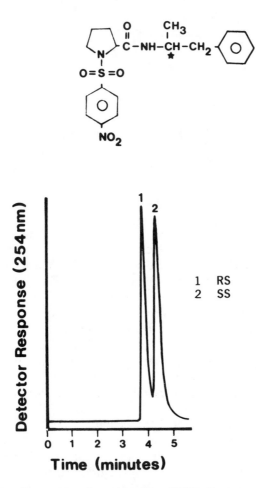

Figure 11.10. Separation of amphetamine-NPSP diastereomers on Supelcosil LC-Si in chloroform: heptane, 8:2. Reprinted with permission from C. R. Clark and J. M. Barksdale, *Anal. Chem.* **1983,** *56,* 958. Copyright 1983, American Chemical Society.

component in the mobile phase and interact differently with the achiral stationary phase, although the mechanism is not at all clear. Typical intermolecular associations that seem to form the basis of chiral separations are hydrogen bonding, metal chelation, and ion-pair formation. One example is the separation of acid enantiomers by forming ion pairs with chiral bases like quinines.[36] Most of the ligand exchange methods in the two reviews mentioned earlier fall into this category.[31,32]

The use of chiral mobile phases is not well suited for preparative LC; often the chiral agents cannot be removed by volatilization for easy recovery of the analytes.

Chiral Stationary Phases

The use of chiral stationary phases (CSPs) is the most popular method for achieving chiral separations because chiral groups can be bonded to the surface of a stationary support using the technologies described earlier, making LC analysis easy. Pirkle[37] has discussed the various types of CSPs, giving examples and describing how they work, and a recent review of chiral stationary phases[38] divides them into five groups:

1. "Brush"-type bonded phases of the type pioneered by Pirkle.
2. Phases containing "cavities" of which the cyclodextrins are typical.
3. Phases containing helical polymers like cellulose.
4. Phases for ligand exchange.
5. Phases for affinity chromatography.

Types 1 and 2 are the most popular and will be described further.

In his 1984 publication, Pirkle[39] summarized his work synthesizing chiral stationary phases. One of his most successful phases was formed by reacting the silyl propyl amine bonded phase with a chiral N-(3,5-dinitrobenzoly)amino acid to form chiral phases of the general formula shown in Figure 11.11. The linkage (Y) between the reacting groups can be ionic or covalent, as shown, and both types are now commercially available and are known as *Pirkle columns*. The ionic CSPs can be used

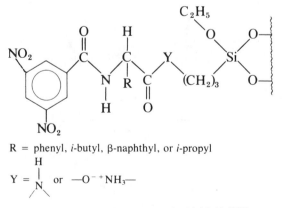

R = phenyl, *i*-butyl, β-naphthyl, or *i*-propyl

$$Y = \begin{array}{c} H \\ | \\ N \\ / \quad \backslash \end{array} \quad \text{or} \quad -O^{-\,+}NH_3-$$

Figure 11.11. Structure of "Pirkle" CSP.

only with relatively nonpolar mobile phases, so the covalent CSPs were designed for use with polar mobile phases. The amino acid moieties used in the commercial products include D-phenylglycine and L-leucine. Manufacturers of these and other CSPs are listed in the 1986 review,[38] which reports the information for eight companies.

Pirkle CSPs have been used for LC separations of the enantiomers of alcohols, sulfoxides, bi-β-naphthols, β-hydroxysulfides, heterocycles such as hydantoins, succinimides, and agents related to propranolol. These phases have also been used in SFC for the separation of phosphine oxide enantiomers.[40]

The second type of chiral separation is based on size exclusion as well as chiral interactions. For example, inclusion complexes are formed between cyclodextrin and molecules with the correct size and shape to fit inside it. Beta-cyclodextrin, with a molecular weight of about 1,000 and with 35 chiral centers is the best one, and the model shown in Figure 11.12 has been proposed to show its shape and interaction with analytes. It appears that the analyte molecule must also have a functional group in the right position to hydrogen bond with a secondary hydroxyl group on the edge of the cage. Computer-generated images of some inclusion compounds illustrate these principles.[41]

Some of the enantiometric compounds separated by LC on a β-cyclodextrin column only 10 cm long are listed in Table 4,[42] and a large number of drug stereoisomer separations have been summarized re-

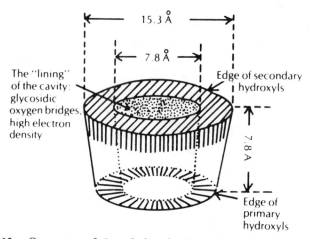

Figure 11.12. Geometry of β-cyclodextrin. Reprinted with permission from *American Laboratory* **1985**, *17*(5), 80. Copyright 1985 by International Scientific Communications.

TABLE 4 Separation Data for Enantiomeric Compounds Using a 10-cm β-Cyclodextrin Bonded Phase Column[a]

Compound	k	α	R_s	Mobile Phase[b]
1. L-Alanine β-naphthylamide	5.1	1.20	2.0	50:50
2. D-Alanine β-naphthylamide	6.1			50:50
3. L-Methionine β-naphthylamide	2.7	1.33	2.4	50:50
4. D-Methionine β-naphthylamide	3.6			50:50
5. L-Alanine β-naphthyl ester	1.0	1.80	2.6	50:50
6. D-Alanine β-naphthyl ester	1.8			50:50
7. Dansyl-L-phenylalanine	3.1	1.23	1.1	55:45
8. Dansyl-D-phenylalanine	3.8			55:45
9. Dansyl-L-leucine	3.0	1.40	2.4	50:50
10. Dansyl-D-leucine	4.2			50:50
11. N-Benzoyl-L-arginine-β-naphthylamide	28.5	1.07	0.6	50:50
12. N-Benzoyl-D-arginine-β-naphthylamide	30.4			50:50
13. (−)Mephobarbital[c]	14.8	1.14	1.6	20:80
14. (+)Mephobarbital[c]	16.9			20:80
15. (−)2,3-O-Isopropylidene-2,3-dihydroxy-1,4-bis(diphenylphosphino)butane	10.56	1.12	1.2	48:52
16. (+)2,3-O-Isopropylidene-2,3-dihydroxy-1,4-bis(diphenylphosphino)butane	11.84			48:52

[a] From Armstrong and DeMond.[42]
[b] Numbers represent the volume percentage of methanol to water.
[c] A 25-cm β-cyclodextrin column was used to generate these data.

cently.[41] Both enantiomers and diastereomers are reported using typical reverse phase mobile phases. Further discussion of the use of β-cyclodextrins in chiral LC separations can be found in references 43 and 44.

A special cyclodextrin bonded phase, with an appropriate binder, has been used in TLC plates to separate isomers, including optical isomers.[45] Another type of TLC plate made by Macherey-Nagel uses a reverse phase bonded layer with Cu^{2+} ions and an unspecified chiral reagent.

Choosing a CSP. Exactly how a CSP works is not known, and the choice of a CSP for a particular separation has been mostly by trial and error. However, some of the principles of *chiral recognition* have been identified for the Pirkle columns.

Pirkle[37] has observed that three simultaneous interactions between the chiral stationary phase and the analytes must occur for chiral recognition and separation. Typical interactions are those discussed in Chapter 3 and include hydrogen bonding, dipole–dipole interactions, and dipole–induced dipole interactions. The geometric arrangement of the chiral phase and the analyte must also be suitable, thus imposing a steric requirement. The stationary phase is chosen so that it will interact with the analytes to be separated according to usual expectations of chemical interactions.

Furthermore, chiral recognition is reciprocal. For example, the dinitrobenzoyl amino acid CSP can separate N-acetylated α-arylalkylamines. Therefore, a CSP made with an α-arylalkylamine functionality should be able to separate analytes containing the dinitrobenzoyl functionality. Such has been shown to be the case, and dinitrobenzoyl derivatives of amines, α-amino esters, and amino alcohols have been separated.[36]

Pirkle[37] has also shown that the spacing between the silica surface and the chiral moiety can be important. A change in the number of carbons in the chain attached to the chiral moiety (shown in Figure 11.11 as three carbons) can reverse the order of elution of two enantiomers.

A theoretical approach to the design of CSPs has been started using molecular orbital theory and computer simulation.[46] It is hoped that this approach will provide the basis for designing new phases for particular analyses.

Souter has included over 800 references in his 1985 book on chiral separations by GC and LC.[47]

DERIVATIZATION

There are many reasons for performing a chemical reaction on a sample to form derivatives. Two reasons can be identified as beneficial for chromatographic analysis: the derivatization improves either the analysis or the detectability. In many cases the reasons are related to qualitative analysis, and Chapter 6 should be consulted for the more general discussion provided there.

Reasons for Derivatization

The major method for improving an analysis by derivatization is by forming volatile compounds from nonvolatile samples for analysis by GC. Often the derivatives will also be more stable thermally, which is necessary for GC analysis. Examples of these reactions are given later. A wide variety of other reactions have been reported whereby the derivatized analytes are more easily separated than the original ones. One major example is the formation of diastereoisomers, which makes possible the resolution of enantiomers, as discussed earlier in this chapter.

Improved detectivity usually arises from the incorporation of a *chromophore* into the analytes. This is most often used in LC, where the derivatives are designed so that they will have a UV absorption or will fluoresce. In GC, an analogous example is the incorporation into the analytes of functional groups that will enhance their detectivity by a selective detector such as the ECD. The purpose of forming the derivatives is to improve the limit of detection or the selectivity or both. Another example is the use of deuterated reagents to form derivatives that can be easily distinguished by their higher molecular weight when analyzed by GC/MS or LC/MS.

Methods

The methods of derivatization can be divided into two categories: pre- and postcolumn methods and off-line and on-line methods. For example, the formation of volatile derivatives for GC is usually prepared off-line in separate vials before injection into the chromatograph (precolumn). There are a few exceptions where the reagents are mixed and injected together; the derivatization reaction occurs in the hot GC injection port (on-line). Precolumn reactions that do not go to completion will produce mixtures that are more complex than the starting sample. As a result, excess reagent is usually used to drive the reaction to completion, thus leaving an excess of the reagent in the sample. Unless a prior separation step is used, the chromatographic method must be designed to separate these additional impurities. When performed off-line, the precolumn reactions can be used with slow reactions and can be heated.

The postcolumn methods usually provide detectivity and are run on-line. The classic example, which has been mentioned several times, is the ninhydrin reaction with amino acids. For best results, the reaction must be fast and the mixing chamber efficient without introducing excessive dead volume. Most of the examples are in LC.

In the United States, two companies have specialized in supplying derivatization reagents, and they provide considerable information about derivatization in their literature.[48,49] The examples are so numerous that several books have been published on the subject[50–54]; the following discussion will only highlight the major uses.

Examples

The two major examples of derivatization in chromatography are the formation of volatiles for GC and the creation of UV detectivity or fluorescence in LC. Each will be discussed briefly.

Volatile Derivatives for GC. The reactions to produce volatile derivatives can be classified as silylation, acylation, alkylation, and coordination complexation. Examples of the first three types are included in Table 5

TABLE 5   Guide to Derivatization^a

Functional Group	Method	Derivatives
Acids	Silylation Alkylation	$RCOOSi(CH_3)_3$ $RCOOR'$
Alcohols and phenols— unhindered and moderately hindered	Silylation Acylation Alkylation	$R—O—Si(CH_3)_3$ $R—O—\overset{\displaystyle O}{\overset{\|}{C}}—PFA$ $R—O—R'$
Alcohols and phenols— highly hindered	Silylation Acylation Alkylation	$R—O—Si(CH_3)_3$ $R—O—\overset{\displaystyle O}{\overset{\|}{C}}—PFA$ $R—O—R'$
Amines (1° & 2°)	Silylation Acylation Alkylation	$R—N—Si(CH_3)_3$ $R—N—\overset{\displaystyle O}{\overset{\|}{C}}—PFA$ $R—N—R'$
Amines (3°)	Alkylation	PFB Carbamate
Amides	Silylation (a) Acylation (b) Alkylation (c)	(a) $R\overset{\displaystyle O}{\overset{\|}{C}}—NHSi(CH_3)_3$ (unstable) (b) $R\overset{\displaystyle O}{\overset{\|}{C}}—NH—\overset{\displaystyle O}{\overset{\|}{C}} PFA$ (c) $R\overset{\displaystyle O}{\overset{\|}{C}}—NHCH_3$
Amino acids	Esterification/Acylation Silylation (a) Acylation + Silylation (b) Alkylation (c)	(a) $\underset{\underset{N-Si (CH_3)_3}{\|}}{RCHCOOSi(CH_3)_3}$ (b) $\underset{\underset{N—TFA}{\|}}{RCHOOSi(CH_3)_3}$ (c) $\underset{\underset{NHR'}{\|}}{RCHCOOR'}$
Catecholamines	Acylation + Silylation (a) Acylation (b)	(a) (b)

TABLE 5   Guide to Derivatization^a (*continued*)

Functional Group	Method	Derivatives

Carbohydrates and sugars	Silylation (a)
	Acylation (b)
	Alkylation (c)
		(a) O Si(CH$_3$)$_3$
		         |
		       —(CH$_2$)$_x$—
		
		(b) OTFA
		         |
		       —(CH$_2$)$_x$$^-$
		
		(c) OR
		         |
		       —(CH$_2$)$_x$$^-$

Carbonyls	Silylation	TMS—O—N=C⟨
	Alkylation	CH$_3$—O—N=C⟨

^{a} Courtesy of Regis Chemical.
^{b} Abbreviations: TMS = Trimethyl silyl; PFA = Perfluoroacyl; TFA = Trifluoracetyl; HFB = Heptafluorobutyryl.

and include the organic functional groups: hydroxyl, amine, and carbonyl. The fourth reaction type, coordination complexation, is used with metals, and typical reagents are trifluoroacetylacetone and hexafluoroacetylacetone.[55] Drozd[56] has reviewed the field and provided over 600 references.

Silylation reactions are very popular and need further description. A variety of reagents are commercially available, and most are designed to introduce the trimethylsilyl group into the analyte to make it volatile. A typical reaction is the one between bis-trimethylsilylacetamide (BSA) and an alcohol:

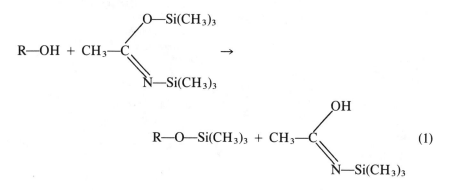

$$R—O—Si(CH_3)_3 + CH_3—C \quad (1)$$

A closely related reagent contains the trifluoroacetamide group and produces a more volatile reaction by-product (not a more volatile derivative); the reagent is bis(trimethylsilyl)-trifluoroacetamide (BSTFA).

If a solvent is used, it is usually a polar one, and the bases DMF and pyridine are common. An acid catalyst such as trimethylchlorosilane (TMCS) and heating are sometimes needed to speed up the reaction. In general, the ease of reaction follows the order[52]:

alcohols ≥ phenols ≥ carboxylic acids ≥ amines ≥ amides

The order of reactivity of the reagents is

TMSIM ≥ BSTFA ≥ BSA ≥ MSTFA ≥ TMSDMA ≥ TMSDEA ≥
MSTA ≥ TMCS ≥ HMDS

(See one of the references for the actual chemical names.)

TABLE 6   Derivatives for LC[a]

Analyte Functional Group	Reagents for	
	UV Absorption	Fluorescence
Acids, carboxylic	O-p-Nitrobenzyl-N, N'-diisopropyliso- urea (PNBDI) p-Bromophenacyl bromide (PBPB)	4-Bromomethyl-7-methoxycoumarin
Alcohols	3,5-Dinitrobenzoyl chloride (DNBC)	—
Aldehydes	p-Nitrobenzyloxy- amine HCl (PNBA)	Dansyl hydrazine
Amines (1° & 2°)	N-Succinimidyl- p-nitrophenyl acetate (SNPA)	7-Chloro-4-nitrobenzo-2-oxa-1,3- diazole (NBD chloride)
	DNBC	Dansyl chloride o-Phthaladehyde (OPT)
Amino acids	SNPA	OPT
Isocyanates	p-Nitrobenzyl-N-n- propylamine HCl	—
Ketones	PNBA	Dansyl hydrazine
Phenols	DNBC	Dansyl chloride
Thiols	DNBC	NBD chloride

[a] Regis Chemical; used with permission.

Derivatives for LC Detection. The lack of a universal detector for LC and the popularity of the UV detector have caused chromatographers to seek derivatization reactions that introduce UV chromophores into sample analytes. In those instances where the derivative also fluoresces, additional sensitivity can be obtained by fluorometric detection. Table 6 contains a list of the most common derivatizing reagents for this purpose.

More details can be found in a recent review[57] and a book on the subject.[58]

REFERENCES

1. D. R. Gere, *Science*, **1983**, *222*, 253.
2. *J. Chromatogr. Sci.* **1986**, *24*, 226–258.
3. D. W. Later, B. E. Richter, W. D. Felix, M. R. Andersen, and D. E. Knowles, *Am. Lab.* **1986**, *18*(8), 108.
4. J. C. Fjeldsted and M. L. Lee, *Anal. Chem.* **1984**, *56*, 619A.
5. M. G. Rawdon and T. A. Norris, *Am. Lab.* **1984**, *16*(5), 17.
6. V. Berry, *LC/GC Mag.* **1986**, *4*, 470.
7. S. R. Springston, P. David, J. Steger, and M. Novotny, *Anal. Chem.* **1986**, *58*, 997.
8. P. A. Peaden and M. L. Lee, *J. Chromatogr.* **1983**, *259*, 1.
9. R. E. Clement, F. I. Onuska, F. J. Yang, G. A. Eiceman, and H. H. Hill, Jr., *Anal. Chem.* **1986**, *58*, 321R. Contains 42 references on SFC.
10. M. G. Rawdon, *Anal. Chem.* **1984**, *56*, 831.
11. J. C. Fjeldsted, R. C. Kong, and M. L. Lee, *J. Chromatogr.* **1983**, *279*, 449.
12. R. D. Smith, J. L. Fulton, R. C. Petersen, A. J. Kopriva, and B. W. Wright, *Anal. Chem.* **1986**, *58*, 2057.
13. S. L. Smith, *J. Chromatogr. Sci.* **1984**, *22*, 143.
14. See, e.g., W. H. McFadden, *J. Chromatogr. Sci.* **1979**, *17*, 2.
15. C. A. Cramers, G. J. Scherpenzeel, and P. A. Leclercq, *J. Chromatogr.* **1981**, *203*, 207.
16. P. R. Griffiths, J. A. de Haseth, and L. V. Azarraga, *Anal. Chem.* **1983**, *55*, 1361A.
17. J. F. Schreider, J. C. Demirian, and J. C. Stickler, *J. Chromatogr. Sci.* **1986**, *24*, 330.
18. F. W. Karasek and A. C. Viau, *J. Chem. Educ.* **1984**, *61*, A233.
19. A. L. Burlinghame, T. A. Baillie, and P. J. Derrick, *Anal. Chem.* **1986**, *58*, 165R.
20. R. S. McDonald, *Anal. Chem.* **1986**, *58*, 1906.
21. C. L. Wilkins, *Science* **1983**, *222*, 291.

22. A. P. Bruins, *J. Chromatogr.* **1985**, *323,* 99.

23. R. E. Majors, *LC Mag.* **1983**, *1,* 488.

24. C. R. Blakley and M. L. Vestal, *Anal. Chem.* **1983**, *55,* 750; T. Covey and J. Henion, *Anal. Chem.* **1983**, 2275; **1984**, *56,* 2.

25. T. R. Covey, E. D. Lee, A. P. Bruins, and J. D. Henion, *Anal. Chem.* **1986**, *58,* 1451A.

26. J. B. Crowther and J. D. Henion, *Anal. Chem.* **1985**, *57,* 2711.

27. R. D. Smith and H. R. Udseth, *Anal. Chem.* **1983**, *55,* 2266; R. D. Smith, H. R. Udseth, and H. T. Kalinoski, *Anal. Chem.* **1984**, *56,* 2971.

28. R. D. Smith, B. W. Wright, and H. R. Udseth, in *Chromatography and Separation Chemistry,* S. Ahuja (ed.), American Chemical Society, Washington, D.C., 1986, pp. 260–293.

29. E. Gil-Av and D. Nurok, *Adv. Chromatogr. N. Y.* **1974**, *10,* 99.

30. I. S. Krull, *Adv. Chromatogr. N. Y.* **1978**, *16,* 175.

31. V. A. Davankov, *Adv. Chromatogr. N. Y.* **1980**, *18,* 139.

32. V. A. Davankov, A. A. Kurganov, and A. S. Bochkov, *Adv. Chromatogr. N. Y.* **1983**, *22,* 71.

33. C. R. Clark and J. M. Barksdale, *Anal. Chem.* **1984**, *56,* 958.

34. R. H. Buck and K. Krummen, *J. Chromatogr.* **1984**, *315,* 279.

35. C. H. Lochmuller and R. W. Souter, *J. Chromatogr.* **1975**, *113,* 283.

36. C. Pettersson, *J. Chromatogr.* **1984**, *316,* 553.

37. W. H. Pirkle, in *Chromatography and Separation Chemistry,* S. Ahuja (ed.), ACS Symposiuum Series 297, American Chemical Society, Washington, D.C., 1986.

38. R. Dappen, H. Arm, and V. R. Meyer, *J. Chromatogr.* **1986**, *373,* 1.

39. W. H. Pirkle, M. H. Hyun, and B. Bank, *J. Chromatogr.* **1984**, *316,* 585.

40. P. A. Mourier, E. Eliot, M. H. Caude, R. H. Rosset, and A. G. Tambute, *Anal. Chem.* **1985**, *57,* 2819.

41. D. W. Armstrong, T. J. Ward, R. D. Armstrong, and T. E. Beesley, *Science* **1986**, *232,* 1132.

42. D. W. Armstrong and W. De Mond, *J. Chromatogr. Sci.* **1984**, *22,* 411.

43. T. E. Beesley, *Am. Lab.* **1985**, *17*(5), 78.

44. W. L. Hinze, T. E. Riehl, D. A. Armstrong, W. De Mond, A. Alak, and T. Ward, *Anal. Chem.* **1985**, *57,* 237.

45. A. Alak and D. W. Armstrong, *Anal. Chem.* **1986**, *58,* 582.

46. K. Lipkowitz, J. M. Landwer, and T. Darden, *Anal. Chem.* **1986**, *58,* 1611.

47. R.. W. Souter, *Chromatographic Separations of Stereoisomers,* CRC Press, Boca Raton, Fla., 1985.

48. *Introduction to Derivatization,* Regis Chemical, Morton Grove, Ill.

49. *Handbook of Derivatization,* Pierce Chemical, Rockford, Ill.

50. A. E. Pierce, *Silylation of Organic Compounds,* Pierce Chemical, Rockford, Ill., 1968.

51. J. F. Lawrence and R. W. Frei, *Chemical Derivatization in Liquid Chromatography,* Elsevier, Amsterdam, 1976.

52. K. Blau and G. S. King (eds.), *Handbook of Derivatives for Chromatography,* Heyden, London, 1978.

53. D. R. Knapp, *Handbook of Analytical Derivatization Reactions,* Wiley, New York, 1979.

54. J. Drozd, *Chemical Derivatization in Gas Chromatography,* Elsevier, Amsterdam, 1981.

55. R. W. Mosier and R. E. Sievers, *Gas Chromatography of Metal Chelates,* Pergamon Press, Oxford, 1965.

56. J. Drozd, *J. Chromatogr.* **1975,** *113,* 303.

57. R. W. Frei, H. Jansen, and U. A. Th. Brinkman, *Anal. Chem.* **1985,** *57,* 1529A.

58. I. S. Krull (ed.), *Reaction Detection in Liquid Chromatography,* Dekker, New York, 1986.

SELECTION OF A METHOD 12

This chapter is intended to provide some suggestions for the novice in chromatography who is charged with the responsibility of devising a method of chromatographic analysis for a given problem. It assumes that a decision has already been made that chromatography is the best choice for the analysis, and it does not compare chromatography with competing methods. It suggests a method of attack for a separation problem and compares the various forms of chromatography in choosing a method.

METHODS OF ATTACK

There are three ways to attack a separation problem: by experimentation, using the best theoretical predictions like those contained in this monograph; by searching the literature in an organized and thorough way to learn what others have done; and by asking someone who may know. The last suggestion may appear to be facetious, but it is probably the best place to begin. If there is anyone you know who might have had some experience with the type of problem you are trying to solve, he or she may be able to give you more information in the shortest time than you can get by either of the other two modes of action. If you are a student, the instructor probably could give you the information you need, but she/ he may be unwilling to do so if that would defeat the educational value of the experiment.

If you are in industry, however, there may be someone in the same lab or a similar lab within your company to whom you can turn for advice. Too often, resource people within an organization are not consulted and insufficient attention is paid to the transfer of information between labs and departments. Do not waste time reinventing the wheel.

There are a number of other people or organizations to whom you can turn for help. The company from whom who purchased your chromatographic equipment or from whom you plan to purchase equipment may be able to help you. A number of these companies maintain applications laboratories for this purpose. Contact your sales representative.

Several distributors of chromatographic supplies and accessories also have experts in their employ who may be able to help you. Some even publish regular newsletters and product bulletins that they will be willing to send you. If you need the names of such firms, check the directory published annually by *Analytical Chemistry* and other journals.

Literature Searching

Searching the literature is much easier and faster than it used to be since the necessary data bases are stored on computer. For example, a complete literature search of *Chemical Abstracts* back to 1967 can be made by computer, and it will cover most of the important work in GC and LC. Other data bases are available, including *Analytical Abstracts* and others specifically in chromatography; several of the latter will be mentioned later. It is highly likely that you will find some references for almost any application because chromatography is a mature science. Indeed, the problem may be to select the best references from among a large number and to choose the best method from a variety of possibilities. If the retrieved references are old, the modern principles and methods described in this monograph should be applied. For example, a GC method using a packed column could probably be improved by substituting an OT column.

There are other alternatives for searching the literature if a complete computerized search is not possible or desirable. Ever since the early days of GC, beginning in 1958, the GC Discussion Group in London has published quarterly abstracts of GC papers during the year, ending with a comprehensive annual index. In 1970 the abstracting service was extended to LC, and the name of the group has been changed to the Chromatography Society to reflect the added coverage. The abstracts have been combined into annual volumes and are available from Elsevier Applied Science Publishers.[1]

Similar in coverage and format is the abstracting service available from Preston Publications.[2] They began GC abstracts in 1958 (including all references back to 1952) and LC abstracts in 1969. Both data bases are published monthly in hard copy, and the GC literature can be searched by computer.

Surveys of the GC and LC literature have also been published as multivolume sets covering the literature chronologically.[3,4] While these may not be as easy to use as the computerized abstracts, they do indicate how complete the coverage of chromatographic literature is.

Several journals provide reviews and bibliographic data. *Analytical Chemistry*[5] publishes biennial reviews in spring issues and currently covers chromatography with three reviews: GC, column LC, and thin-layer and paper chromatography. The coverage and style depends on the authors for a given year or topic, but the wide availability of this journal makes these reviews an excellent source of information on the developments in chromatography. In the alternate years the journal's reviews cover the different areas of application, and these too are a good source of information.

Reviews in the *Journal of Chromatography* are on specific subtopics, and they used to be published separately as *Chromatographic Reviews*. They will not be as useful in searching for a method of analysis. However, the journal also publishes bibliographic data in the three areas—GC, LC, and planar chromatography (plus electrophoresis). The most recent bibliographies are volume 335 (1985–1986) and volume 304 (1984).

There are several compilations of chromatographic data that may be of use in setting up a method of analysis. The ASTM has published retention data for GC,[6] LC,[7] and SEC.[8]

McReynolds[9] has published GC retention data for 400 compounds on 77 stationary phases at two temperatures. These data are of particular value because they were obtained under the same conditions by the same person and are therefore self-consistent. For TLC data, see the references provided at the end of Chapter 10.

The Experimental Approach

It is unlikely that the literature will fail to reveal some information that can provide a clue for the initial experiments. However, if that were to be the case, and if there were no information available about the sample, how would one proceed? Let us assume that a separation is needed and that it has been decided that the best possible separation would be by chromatography.

Scott[10] was the first to suggest that there are three objectives that need to be considered in a chromatographic analysis: speed, resolution, and quantity of sample to be separated. Since these three objectives are interrelated, improvements in one are achieved at the expense of the others, and their relationship is best depicted as a triangle, as shown in Figure 12.1. As a rough guide to method selection, the triangle is divided into sections showing the most likely method for achieving the desired performance with respect to the three objectives. The dividing lines are arbitrary, of course, and this approach is oversimplified, but it does give an overview of the most common uses of the various chromatographic modes.

Probably the most important first step in designing a particular analysis would be to determine the volatility of the sample to ascertain if it can be run by GC. A microdistillation should be run to provide this information. Short, lightly loaded OT columns can handle samples with boiling

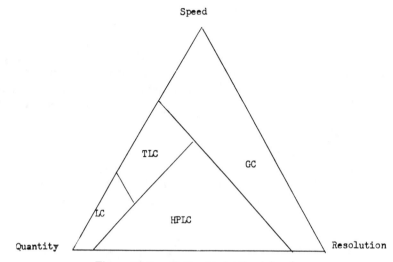

Figure 12.1. Method selection triangle.

points up to about 500°C or above. In general, GC is the preferred method for fast, high resolution work, and it should be the method selected if at all possible.

It would also be helpful to run an infrared spectrum of the sample to determine what functional groups are present. This will help in the selection of a stationary phase. Cooling the sample can be easily accomplished, and, if any solids crystallize, they can be easily removed, thus providing a simple separation. Remember that freezing points are not necessarily related to boiling points, and they often represent a simple way to clean up a complex sample.

Volatile Samples. If it has been determined that the sample is volatile enough to be run by GC, an OT column should be chosen to match the sample as closely as possible; that is, a nonpolar column like DB-1 should be chosen for a nonpolar sample, and a polar column for a polar sample. For initial screening, a short length of 10 to 15 m is desirable, as is a light load of about 0.25 μm. A rather fast temperature program of perhaps 10°/min should be used, starting at a temperature below the lowest boiling component in the sample and proceeding to the highest boiler or to the upper temperature limit of the column (whichever is lower). The upper temperature should be maintained until it can be assumed that all analytes have been eluted.

Depending on the quality of the separation obtained on the fast screening run, temperatures and/or stationary phases can be selected to improve the analysis.

Nonvolatile Samples. Figure 12.2 shows one flow sheet for selecting a chromatographic method. Normally nonvolatile samples are run by LC, but consideration should be given to the possibility of derivatizing the sample to get volatility adequate for GC. (Consult Chapter 11 for possible derivatizations.) Other variables considered in Figure 12.2 are the number of samples and the importance of speed.

The analysis of complex samples can probably be facilitated if they are cleaned up prior to analysis by using short columns and/or extractions, as described in Chapter 9. Flash chromatography (Chapter 9) has become very popular for fast, crude preliminary separations. Other preliminary

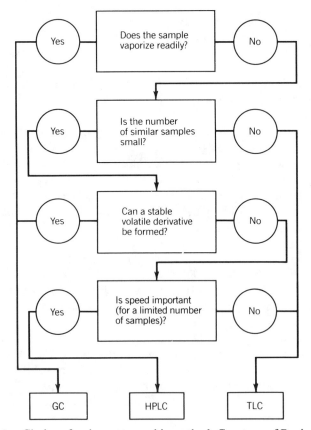

Figure 12.2.   Choice of a chromatographic method. Courtesy of Regis Chemical.

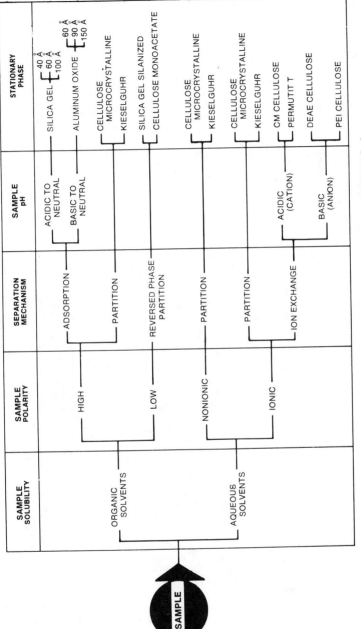

Figure 12.3. Stationary phase selection guide. Courtesy of EM Labs.

281

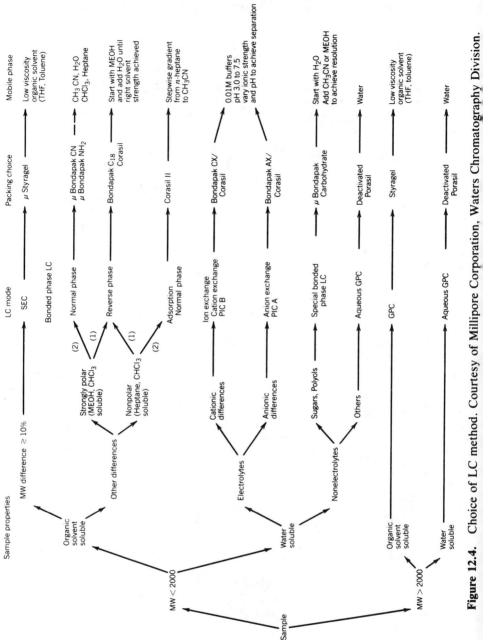

Figure 12.4. Choice of LC method. Courtesy of Millipore Corporation, Waters Chromatography Division.

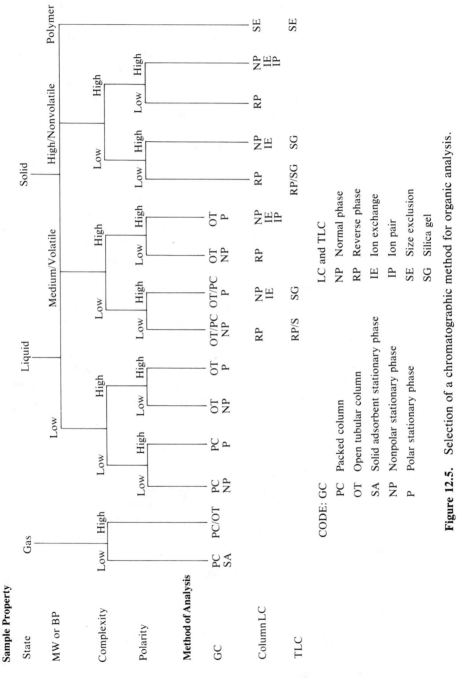

Figure 12.5. Selection of a chromatographic method for organic analysis.

Sample Property

State

MW or BP

Complexity

Polarity

Method of Analysis

GC

Column LC

TLC

CODE: GC

PC   Packed column

OT   Open tubular column

SA   Solid adsorbent stationary phase

NP   Nonpolar stationary phase

P   Polar stationary phase

LC and TLC

NP   Normal phase

RP   Reverse phase

IE   Ion exchange

IP   Ion pair

SE   Size exclusion

SG   Silica gel

283

treatments to be considered are drying to remove water, derivatization, pH adjustment, and preconcentration.

Fast screening is often accomplished by TLC, as indicated in Figure 12.2. The choice of stationary phase can be based on sample solubility and polarity according to Figure 12.3. Silica gel is commonly used for a test run, and homemade silica gel layers on microscope slides (see Chapter 10) provide fast and inexpensive screening. A polar mobile phase is used first, followed by less polar ones and mixtures, depending on the chromatographic results.

If there is any reason to suspect that the sample contains analytes with a wide range of molecular weights or any analytes with molecular weights over 2,000, then SEC should be considered. Figure 12.4 is typical of the flow diagrams provided by many manufacturers of LC columns and column packings. It is based on size or molecular weight, sample solubility, and sample polarity. Obviously there are many choices, and the selection process is not simple. Gradient elution would facilitate screening by LC. Chapter 10 contains extensive directions for selecting the best LC mobile phase.

Figure 12.5 contains a summary of the selection process based on the following sample properties: state, molecular weight or boiling point, complexity, and polarity. It includes gas samples that have been excluded from prior discussion; obviously they would be run by GC. For some sample types, both GC and LC methods are possible, but the figure does not include derivatization. This figure is only a rough guide; many exceptions are likely. Still, it does suggest a beginning point for an analysis, and it does summarize the principles we have been discussing.

SUMMARY

In most cases a combination of approaches is probably desirable—some discussion with colleagues, some literature searching, and some experimentation. Personal preference and availability of instrumentation are strong factors in the decision making process, and they have not been included in this discussion. In this regard, the author's preferences undoubtedly influenced the material presented, but it is hoped, only minimally.

REFERENCES

1. *Chromatography Abstracts*, Elsevier Applied Science Publishers, Barking, England; in North America, Elsevier Science Publishing, New York.

2. *GC and LC Literature: Abstracts and Index*, Preston Publications, Niles, Ill.

3. A. V. Signeur, *Guide to Gas Chromatography Literature*, Plenum, New York, 1979; Vol. 1, 1964, Vol. 2, 1967, Vol. 3, 1974, Vol. 4, 1979.

4. H. Colin, A. M. Krstulovic, J. Excoffier, and G. Guiochon, *A Guide to the HPLC Literature*, Vol. 1 (1966–1979), Vol. 2 (1980–1981), and Vol. 3 (1982), Wiley, New York.

5. The latest reviews are in *Anal. Chem.* **1986,** *58.*

6. O. E. Schupp and J. E. Lewis (eds.), *Compilation of Gas Chromatographic Data*, 2d ed., ASTM Special Publication AMD-25A and *Supplement 1*, Publication AMD-25A S-1, American Society for Testing and Materials, Philadelphia, 1967 and 1971.

7. *Liquid Chromatographic Data Compilation*, ASTM Publication AMD-41, American Society for Testing and Materials, Philadelphia, 1975.

8. *Bibliography on Liquid Exclusion Chromatography*, ASTM AMD-40 and AMD-40-S1, American Society for Testing and Materials, Philadelphia, 1972 and 1975.

9. W. O. McReynolds, *Gas Chromatographic Retention Data*, Preston Technical Abstracts, Evanston, Ill., 1966.

10. R. P. W. Scott, in *Gas Chromatography, 1964*, A. Goldup (ed.), Institute of Petroleum, London, 1965, pp. 25–37.

APPENDIX

TYPICAL CHROMATOGRAPHIC CALCULATIONS

Figure A.1 shows a liquid chromatographic separation of acetylacetonate chelates of beryllium, chromium, ruthenium, and cobalt. The conditions of the separation were:

Stationary phase: a ternary mixture of 64% ethanol, 34% water, and 1.6% isooctane.

Mobile phase: a ternary mixture of 98% isooctane, 2% ethanol, and 0.08% water.

V_M	1.32 mL
V_S	0.142 mL
F	0.291 mL/min
Column length L	23 cm
Inside diameter of column	2.7 mm
Inlet pressure P_i	50 atm
Chart speed	1 cm/min

Calculate β for this column, and for both the beryllium and the cobalt chelates calculate the following parameters: V_R, V_N, k, R_R, K, n, and H. For the unresolved mixture of chromium and ruthenium chelates, calculate the resolution, R_s, assuming that the peaks are of equal height.

The partition coefficients for these chelates have been determined by static methods and found to be as follows: beryllium, 4.4; chromium, 16.1; ruthenium, 23.2; and cobalt, 33.5.

Figure A.2 shows a gas chromatogram of toluene. The operating conditions and some other necessary data are given below:

Column: 48 × 0.25 in. (outside diameter); inside diameter = 5.0 mm; packed with 12.0 g containing 20% (w/w) dinonyl phthalate on 80/100 mesh Chromosorb P

Column temperature	100°C
Ambient temperature	25°C
Ambient pressure P_0	760 torr

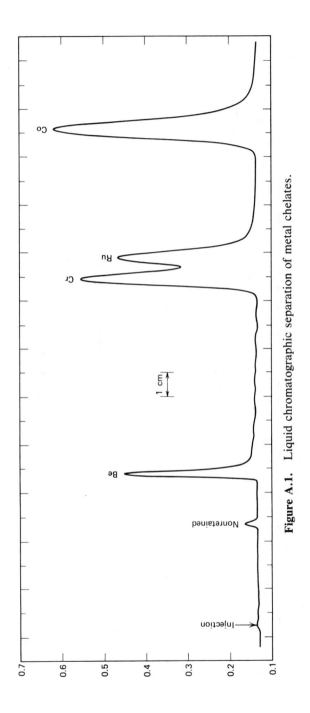

Figure A.1. Liquid chromatographic separation of metal chelates.

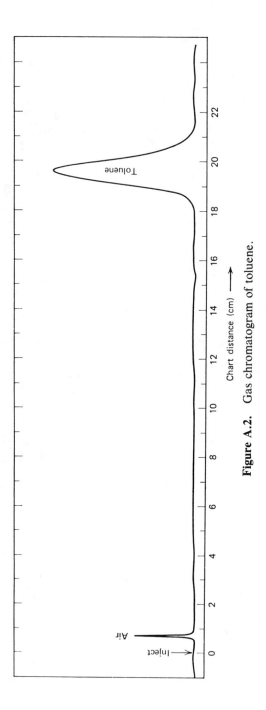

Figure A.2. Gas chromatogram of toluene.

Outlet flow measured at ambient conditions

with a soap-bubble flowmeter	80 mL/min
Chart speed	1 in./min
Density of Chromosorb P	2.26 g/mL
Density of dinonyl phthalate	1.03 g/mL
Vapor pressure of water at 25°C	24 torr
Calculated pressure correction factor j	0.65

Calculate as many chromatographic parameters as you can, including the total volume of the column. The actual calculations follow.

Note that in this example the symbol V_G is used to denote the mobile phase (which is a gas) rather than the general symbol V_M. Similarly, V_L is used rather than V_S for the stationary phase.

CALCULATIONS: GAS CHROMATOGRAM OF TOLUENE

1. Total volume of column, V_t. Radius = 0.25 cm. $V_t = 3.14 \times (0.25)^2 \times 4 \times 12 \times 2.54 = 23.9$ mL.

2. $t_R = 19.45$ cm $\times$ 1 min/in. $\times$ 1/2.54 in./cm = 7.66 min.

3. $t_R = (19.45 - 0.70) \times 1/2.54 = 7.39$ min, or $t_M = 0.70/2.54 = 0.28$ and $t'_R = 7.66 - 0.28 = 7.38$ min.

4. $F_c = 80 \times 373/298 \times 736/760 = 97$ mL/min.
$\overline{F}_c = 97 \times 0.65 = 63$ mL/min.

5. $V_R = 7.66 \times 97 = 743$ mL.

6. $V'_R = 7.39 \times 97 = 716$ mL.

7. $V^0_R = 743 \times 0.65 = 482$ mL.

8. $V_N = j \times V'_R = 716 \times 0.65 = 466$ mL, or $V_N = V^0_R - V^0_G = 482 - 18 = 464$ mL.

9. $V^0_G = t_M \times \overline{F}_c = 0.28 \times 63 = 17.6$ mL.

10. $V_L = 12$ g $\times 0.20 \times 1/1.03$ mL/g = 2.33 mL.

11. $V_{SS} = 12 \times 0.80 \times 1/2.26$ mL/g = 4.25 mL.
$$V^0_G = 17.6 \qquad 72.8\% = \epsilon_T$$
$$V_L = 2.3 \qquad 9.5\%$$
$$V_{SS} = \underline{4.3} \qquad 17.7\%$$
$$\phantom{V_{SS} = } 24.2 = \text{total volume; compare with volume calculated in item 1.}$$

12. $\overline{u} = \dfrac{\overline{F}_c}{A \times \epsilon_T \times 60} = \dfrac{63}{0.182 \times 0.73 \times 60} = 7.88$ cm/sec, or
$\overline{u} \approx L/60t_M = 122/60 \times 0.28 = 7.26$ cm/sec.

13. $\beta = 17.6/2.33 = 7.6$.
14. $K = V_N/V_L = 465/2.33 = 199$.
15. $k = K/\beta = 199/7.6 = 26.2$
    Also note: $V_R^0 = V_G^0(1 + k') = 17.6(27.2) = 479$ mL (compare with item 7).
16. $R_R = V_G^0/V_R^0 = 17.6/482 = 0.0366$, or toluene spends 3.7% of its time in the mobile phase.
    $(1 - R_R) = 0.9634$, or toluene spends 96.34% of its time in the stationary phase.
    $k = \dfrac{1 - R_R}{R_R} = 96.3/3.6 = 26.3$ (compare with item 15).
17. $n = 16(19.45/1.96)^2 = 16(9.93)^2 = 16(99) = 1584$ plates.
18. $H = L/n = 122$ cm/1584 $= 0.077$ cm $= 0.77$ mm.
19. $h = H/d_p = 0.77/0.163 = 4.4$

LIST OF SYMBOLS AND ACRONYMS

A	Area
A_s	Area on the surface of an adsorbent required by a solute
ACN	Acetonitrile
ASTM	American Society for Testing and Materials
BPC	Bonded phase chromatography, in LC
C	Concentration, usually in mol/L
D	Diffusion coefficient; for example D_L for a liquid
d	Distance between two peak maxima
d_c	Diameter of a column
d_f	Film thickness
d_p	Diameter of a particle
$\mathscr{E}$	Free energy of constant V and T
E_a	Surface energy of an adsorbent (see Snyder solvent parameter)
ECD	Electron capture detector
F	Flow rate: F_o, at outlet; F_c, corrected outlet flow; $\overline{F}_c$, average corrected outlet flow
f	Detector response factor
FID	Flame ionization detector
FPD	Flame photometric detector
FTIR	Fourier transform infrared spectroscopy
GC	Gas chromatography
GLC	Gas–liquid chromatography
GSC	Gas–solid chromatography

$\mathcal{G}$	Gibbs free energy at constant P and T
$\mathcal{H}$	Enthalpy
H	Peak dispersivity or plate height (also known as HETP, height equivalent to a theoretical plate)
HPLC	High performance LC; usage not recommended
HPTLC	High performance TLC; usage not recommended
h	Reduced plate height
I	Retention index of Kovats; also ΔI
IEC	Ion exchange chromatography
IPC	Ion pair chromatography
IR	Infrared spectroscopy
j	Pressure correction factor
K	Equilibrium constant or partition coefficient (based on only one species)
k	Partition ratio or capacity factor
L	Length, e.g., column length
LC	Liquid chromatography
LLC	Liquid–liquid chromatography
LSC	Liquid–solid chromatography
MDQ	Minimum detectable quantity; also detectivity
MS	Mass spectrometry
n	Plate number
n_{eff}	Effective plate number
NPLC	Normal phase LC
ODS	Octadecylsilyl
ORM	Overlapping resolution mapping
OT	Open tubular (column)
P'	Polarity index
P	Pressure: P_i, inlet pressure; P_o, outlet pressure
p	Partial pressure
p^0	Equilibrium vapor pressure
PC	Paper chromatography
PTGC	Programmed temperature GC
q	Configuration factor (in Rate equation)
$\mathcal{R}$	Gas constant
R_f	Retardation factor
R_M	Martin retention parameter
R_R	Retention ratio
R_s	Resolution
r_c	Radius of a column
RI	Refractive index (detector)
RPLC	Reverse phase LC

S^0	Adsorption energy of a solute (see Snyder solvent parameter)
SCOT	Support coated open tubular (column)
SEC	Size exclusion chromatography (also erroneously called gel permeation or gel filtration chromatography)
SFC	Supercritical fluid chromatography
SN	Separation number; see TZ
T	Temperature
T'	Significant temperature (in PTGC)
t	Time
TCD	Thermal conductivity detector
THF	Tetrahydrofuran
TF	Tailing factor, a measure of asymmetry
TLC	Thin layer chromatography
t_R	Retention time
t_R'	Adjusted retention time
TZ	Trennzahl number
u	Velocity of the mobile phase; u_f = velocity of solvent front
UV	Ultraviolet, including ultraviolet absorption detector
V	Volume; V_A is the molar volume of solute A
$\overline{V}$	Molal volume
V_a	Adsorbent surface volume (see Snyder solvent parameter)
V_M	Volume of the mobile phase: V_G, volume of gas; V_L, volume of liquid
V_R	Retention volume
V_R'	Adjusted retention volume
V_R^0	Corrected retention volume
V_N	Net retention volume
V_S	Volume of the stationary phase: V_L, volume of liquid; V_S, volume of solid
V_{ss}	Volume of the solid support
v	Velocity of an analyte
w	Width of a peak at the base
w_h	Width of a peak at half-height
w_x, w_s	Weight of unknown (x) or standard (s)
WCOT	Wall coated open tubular (column)
X	Mole fraction
α	Separation factor, or column selectivity, or relative retention
β	Phase volume ratio
γ	Surface tension

γ_A	Activity coefficient for solute A
δ	Hildebrand solubility parameter
ϵ	Porosity; ϵ_T is total porosity; ϵ_I is interstitial porosity
ϵ^0	Snyder solvent parameter
η	Viscosity
θ	Angle, e.g., in TLC
κ	Permeability
λ	Packing characteristic (in rate equation)
μ	Dipole monent
ν	Reduced velocity
ρ	Polarizability
σ	Standard deviation or quarter-zone width
τ	Detector time constant
ϕ	Volume fraction
Φ	Flow resistance parameter
ψ	Obstruction factor (rate equation)
ω	Packing factor (rate equation)

INDEX

Absorption, 6, 8, 40
Accuracy, definition of, 99
Activity coefficient, 109–110
Adjusted retention volume, 10
Adsorption, 6, 8, 40
Affinity chromatography, 224–225
Area, measurement of, 102
Area normalization, 104
Asymmetry of peaks, 14, 43

Band, definition of, 69
Bonded phase chromatography (BPC),
    166–171
Brockmann activity, 238

Capacity factor, 11
Capillary columns, 110, 118–121
Carbon dioxide phase diagram, 65, 251,
    252
Carbowax, 133
Carrier gases, effect on efficiency in OT
    columns, 37
Centrichromatography, 74
Charge-transfer, 45
Chemisorption, 41, 42
Chiral separations, 261–268
Chiral stationary phases (CSP's), 265–268
Chromatogram, 8
Chromatography:
    classification of, 5
    definition of, 4–7
Clathrate, 46
Clausius–Clapeyron equation, 144
Column:
    micro-bore, 200
    open tubular (OT), 110, 118–121
    packed, 115
    wide-bore, 121
Column selectivity, 20–22, 77
Comparisons:
    gases to liquids, 60
    GC to LC, 60
    peak height to peak area, 101

peaks to bands, 69
R_f to R_R, 71
Compressibility factor, 61–63
Configuration factor, 33
Countercurrent chromatography, 74
Critical point data, 250
Cryogenic GC, 149
Cyclodextrin, 266

Dead volume, 9, 25, 27
Debye forces, 43, 44
DEGS, 133
Derivatization, 81, 82, 209, 268–273
Detectivity, 96–98
    in GC, 122
    MDC and MDM, 97, 130
    MDQ, 96
Detectors:
    classification of, 88–92
    GC, 121–131
    LC, 204–209
Dexsil, 133, 136
Diffusion, 29
Diffusion coefficient, 31
Dispersion forces, 43, 44, 141
Dispersivity, 35, 53, 70
Displacement chromatography, 6
Drift, 93
Dual detectors (dual channel GC), 82–84

Eddy diffusion, 28, 31, 34
Effective plate number, 16
Einstein equation, 35
Electrochemical detectors in LC, 208
Electron capture detector (ECD), 127–129
Eluent strength function, 49–51, 159, 160–
    162
Elution, gradient, 161, 195
Elution chromatography, 6
Equations, summary table of, 22
End capping, 168
External standard, 105
Extra-column zone broadening, 203